What the Animals Tell Me

WHAT THE ANIMALS TELL ME

Beatrice Lydecker

with Ingrid Yates

HARPER & ROW, PUBLISHERS, San Francisco
Cambridge, Hagerstown, New York, Philadelphia
London, Mexico City, São Paulo, Sydney

 For information address Harper & Row, Publishers, Inc., 10 East 53rd Street, New York, N.Y. 10022. Published simultaneously in Canada by Fitzhenry & Whiteside Limited, Toronto.

First Harper & Row paperback edition published 1982.

Library of Congress Cataloging in Publication Data

Lydecker, Beatrice.
What the animals tell me.

1. Animal communication. 2. Animals, Habits and behavior of.
3. Extrasensory perception. 4. Pets. I. Yates, Ingrid, joint author. II. Title.
QL776.L9 599'.05'9 76-9997
ISBN 0-06-250551-3

---

86 10 9 8 7 6 5 4 3

CONTENTS

PREFACE

A book about animals is usually written by a veterinarian or by a breeder or by someone with a story to tell about his or her pets. I have a great deal of experience with pets, pet owners, and breeders, and I consult with veterinarians throughout the country. However, I am not a veterinarian or a full-time breeder, and so this is a different sort of book about animals.

Like any communication, this book is a sharing experience. I want to share with you what a great many animals have so generously shared with me—feelings, thoughts, direct factual information, memories, and past experiences. Animals are capable of both pleasure and pain and of the so-called human emotions. In fact, they are subject to many psychological problems that plague humans. They can be unhappy, lonely, frustrated, homesick, nervous, full of fears. They can be shy, overly agressive, and angry. They can hurt, both physically and emotionally. They can also be happy and full of life.

Animals are often considered "dumb," but I believe this adjective applies only in one sense of the word, that is, they do not speak the human language. They are often considered "intelligent" in varying degrees, but nearly always in terms of human intelligence and/or obedience to human commands. There is also the fact of "likeness." The monkey, biologically endowed with legs, arms, hands, fingers, and so on, is usually considered more intelligent than some other species such as the crocodile or the giraffe because the monkey, with its hands, fingers, and capacity for mimicry, can imitate man in recognizable ways.

Animals are not "dumb" in any sense of the word. They have a nonverbal intelligence and a nonverbal language which, through the years, I have learned. It is like another language although communicated in silence, with great respect on both sides and also with effort and sympathy. I have learned this language, and this book is to tell you about it, to teach you to use it, and above all to tell you how your pets may feel about a great many matters which directly concern them.

Now I should explain what I am and what I am not, or vice versa.

I am not a veterinarian. I am therefore not officially qualified to diagnose, prescribe, or make medical recommendations. That is for your veterinarian to do. I am also not licensed to tell you to give your pet a particular vitamin or mineral for this or that reason. I do offer suggestions throughout this book, and I give my reasons for doing so. I make recommendations for certain animal conditions—both physical and emotional—which have been tested again and again and have proved successful with my animals and with those of my clients and which have also been recommended by many veterinarians (not necessarily all vets since they are just as capable of controversy as any other profession). These suggestions can be interpreted as *sort of* diagnostic, and in some instances they may be considered prescriptions since they range from vitamin supplements to specific diets and medication for particular ailments. Yet they are offered as advice, and I hope they are taken as that. My advice is always based on experience with my own animals or with those animals I've been called in for consultation and also on a close working relationship with a number of veterinarians throughout the country. My first suggestion, therefore, is to urge you to find a good veterinarian if you don't already have one. This person is as important to your pets as a good doctor is to you and your family. I recommend a close relationship between you and your pets, and your pets with their vet. If there is any doubt about an ailment or diagnosis, a good vet will suggest you get another opinion, and possibly a third.

If I am not a veterinarian or a full-time breeder, what am I? And what are my qualifications for writing this book? First and foremost, I am an animal owner and animal lover. The animals I know best are dogs, cats, and horses—purebred, mongrel, for show, for breeding, or companions for home and family. My animal family, needless to say, is large. Over the years it has varied between one and ten dogs, one and three cats, and two horses. On occasion my family has included certain farm animals, ducks, and several cows. I am a part-time breeder of German Shepherds. I may show the puppies or keep them as pets. At this writing my animal family consists of thirteen: three spayed female cats—Snowbunny (alleycat), Tigger Too (Dimpletail Black Manx), Kathy (another alleycat); Blacky, now sixteen years old (Pomeranian); three adult German Shepherds—Princessa (has two pups), five years old, Loverboy (registered Alpha), and Philea, the mother of five pups: Marc Anthony (whom I will keep), Cleopatra, Brutus (Et tu Brutus), Nero (Gabby), and Caesar Augustus. We live in unusual harmony.

My first experience with animals began when I was an infant, so young I can't remember the occasion; I was probably in the cradle, the youngest in a family of eight children. I was born in Islip, a small town in Long Island, New York. When I was five, we moved to a farm not far from Glens Falls, New York. As far back as I can remember, we had house pets, and some farm animals also became house pets. Later, when I was married, my husband and I had several German Shepherds and several cats. We lived on a ranch in Duarte, California, where there was considerable livestock. We had chickens, and we rented pastureland to the local dairy for its pregnant cows. On the ranch I became aware of a certain affinity with animals which I had never experienced before. This led me, over a period of years, into a lot of studying and eventually into a full-time career as animal consultant, or animal analyst. My studies included biology, animal physiology and anatomy, zoology, genetics, principles of psychology, child psychology, nutrition, and food analysis. I have studied with veterinarians but with no intention of ever becoming one; in turn, they consult me on cases that stump them. They know medicine and can prescribe. I can talk to the animal in question, find out about the pain, and suggest a "diagnosis." So I am now considered an expert in animal psychology, and my title, when introduced on a TV show or a lecture platform, is Animal Analyst or Animal Consultant since I consult both with animals and their owners.

Here I should make two points which I will deal with more fully later. First, my work in television with animal trainers responsible for pet-product commercials has made me a strong skeptic on the merits of advertised products. My policy is not to recommend brand-name products. However, in chapter 10 I have broken my rule; I have intentionally warned the reader against certain products I know to be harmful (and possibly fatal) to pets, and I have suggested substitutes that meet the need or the requirement.

Second, the neutering and spaying of my pets is not accidental but a humane action in which I strongly believe. Later I will discuss some myths and misconceptions, some dos and don'ts, so that you can decide what is best for your pet. The same goes for declawing cats that are confined to a house or apartment.

My major talent, however, is not medicine, animal psychology, nutrition, breeding, or training. It is understanding domestic animals by communicating with them. Animals, at least those species I have mentioned, are hidden from us by a language barrier which can cause great difficulties at times and may even result in the animal's death. I don't believe animals like the barrier because, as I gradually became aware of their feelings and began to understand their nonverbal speech, I know they have a lot they want to say.

This book is exactly what the title states: It's what the animals tell me. You may find my experiences incredible; you may raise your eyebrows over some; you may think I'm talking nonsense, or you may want corroboration and more information. In any case, please feel free to write to me. If you have a pet or pets you love and enjoy or if you are worried about an animal, I want to help you help them by being more aware of each other in this wonderful, very special way I have discovered.

I can "look" into a body and see what is wrong with it physically; I can also "see" over great distances. Another part of ESP is the ability to be in a place (walk into a room, perhaps) and "see" what has already happened there. I can see a past event, the surroundings as they were at that time, and the people involved. I believe energies are left behind. This is why people sometimes feel they have been in a particular place before, the experience of *dêja vu,* or even the experience of having lived there in a previous life. In short, reincarnation, another subject about which there is controversy and misunderstanding.

The first part of this book may seem metaphysical or occult. It is not. This section is strictly personal experience; every word is autobiographical. The following four chapters are what might be called the how-to part, the practical day-to-day care, both physiological and emotional, of domesticated pets. I draw on my experience and, of course, on that of the best veterinarians in the country. The last chapter is about my religious beliefs or faith. I do not want my special talents and my ability to talk to and understand animals to be confused with any occult sect or to be considered a supernatural force. I am a practicing Christian, and the special creatures of God's world I love the most are animals.

So my book is part personal adventure and part advice, in varying proportions. Above all, my aim is to share my adventures and discoveries with you and to help you understand your pets and their needs so that both you and they will enjoy life together and find it more rewarding than ever before.

Beatrice Lydecker
North Hollywood, California

ACKNOWLEDGMENTS

I would like to express my appreciation to all the wonderful people who have so graciously allowed me to tell about their animals to illustrate this book. Their generosity will help many other animal owners to establish a better relationship with their pets and to care for them more effectively. My special thanks to many fine veterinarians who have been so helpful with their devotion of time and sharing of their knowledge. Their acceptance of me has made my work so much easier.

I can hardly express the deep gratitude I feel for Dr. M. J. Cain of Cincinnati, Ohio, who has so lovingly and patiently taught me much of what I know. His endorsement of my work has opened many doors to accomplishment for me. He is, in my opinion, a genius in his field, and I count it a privilege to have had the opportunity to learn from a man with his capabilities. My gratitude is also unbounded to Dr. Howard Kurtz of Canoga Park, California, who has been my own trusted veterinarian. His gentleness as well as the wonderful care both he and his staff give to my animals is beautiful to see. My dogs joyfully run to greet them when taken to his hospital for treatment or for boarding. His consulations, along with Dr. Cain's, have helped to make this book what it is.

PART I

A PERSONAL VENTURE

1

AWAKENING

One day, in January 1969, while walking down a street in Monrovia, California, I noticed a German Shepherd dog standing quietly behind the fence of a house. Almost as if he called to me, I went over and petted him through an opening in the gate. Suddenly a heavy sense of rejection and depression came over me. I was always around animals, but this was the first time I had experienced such a feeling, and it frightened me. Most dogs readily adapt to being alone in an outdoor yard, and there seemed no obvious reason for the emotional response I was picking up from him. It baffled me so much that I returned to the house later that day and talked with the owner. My questions surprised him at first, but then he explained. An injury sustained in an automobile accident had confined him at home for several months; and he had bought the dog for companionship for him and his young son. Today he had finally returned to work, and his son had started nursery school. The dog, Duke, was left alone for the first time. Learning this, I was no longer mystified by the feelings I had picked up from Duke.

Shortly thereafter as I stood on a corner one afternoon waiting for the traffic light to change, I felt strangely forlorn. I experienced a soundless cry for help. Clear yet inaudible, its desperation grew in intensity until I could not ignore it. No one was around except a Doberman Pinscher on guard duty behind a warehouse fence. I had seen the dog before, growling at passersby, occasionally even charging at the fence. Still I could not resist going over to the animal. I approached carefully, and to my astonishment the Doberman pranced over to me and nuzzled as far as it could through the wire fence, pressing its nose ever so gently against my outstretched hand. I felt I had actually experienced this animal's emotions.

These were my first encounters with animals on this level of communication, and they accounted for my beginning awareness that I had a strange, unusually perceptive ability. I was thirty-one years old and had worked as a supervisor of a women's prison, a bookkeeper, and a cosmetic sales representative. While working on a forty-acre avocado ranch in Duarte, California, my gift of extrasensory perception revealed itself. I was in charge of the irrigation, and I now believe that living close to the soil for two years opened my senses.

The change was gradual and dreamlike; yet I remember the time clearly. On the ranch during the summer of 1968, a sense of wonder began to awaken in me that replaced many years of restlessness. Purpose and direction seemed to come into my life. I became part of the avocado groves, discovering a kinship with the constant, soft breezes. I experienced a oneness with wind and rain and reacted to the slightest change in atmospheric conditions. My view of the world was shifting. In this very special place, I felt I belonged to nature. I sensed a oneness with God's creation and experienced the gradual unfolding of my extrasensory perception. It's difficult to convey the full impact of this feeling, but it includes exhilarating peace, inner knowledge, and awareness that even greater mysteries are waiting to be experienced; immediate physical limits are transcended. The opening of awareness brings an affection for all life. And that was my beginning.

I was young and felt inexperienced in this kinship with the world around me, but as time passed, I realized how necessary caring for all life is. In fact, it is vital. In order to communicate with animals, in two-way mental communication, a willingness to love all forms of life must be part of the individual. As a basic first step, I surrendered all my misconceptions about animals because they limit and prevent communication. The second step was to recall my experience of oneness with the serenity of nature. By nurturing these feelings I eventually came to see with my mind's eye instead of with just my physical eyes. Then I began to study animals as beings in their own right with expressive, individual personalities. I looked for reactions—no matter how absurd they appeared to my sense of logic. These reactions were signs of the animal's responsiveness.

I remember an experience with Blacky, my Pomeranian who was then eight years old and lived on the ranch with me. One afternoon, I planned a shopping trip to Pasadena but dreaded the prospect of driving several hours through smog and traffic, parking in the terrible heat, and traipsing from store to store. I dawdled in my room while dressing and delayed leaving as long as possible. The familiar noise of the screen door opening and closing told me that Blacky had already headed for the car. He always knew when it was time to go for a ride and could be found patiently sitting at the car with a whatever-took-you-so-long expression on his face. As I was getting ready to leave, I said to myself, *Okay, friend, I know you've planted yourself by the car expecting to go with me, but this time you're in for a surprise because you're staying home. It's too hot for you to sit in the car and wait while I shop.* However, when I reached the car, he was nowhere in sight. I looked in his favorite hiding places until I was tired of searching. This surprising disappearing act added to my growing impatience. I was already late getting started, and I wanted to put him inside before I left. I called, yelled, whistled—still no Blacky. I gave up and drove off. One-quarter mile later, I found a little black puff sitting in the middle of the road waiting to be picked up. I was stunned! Imagine being outwitted by a dog! Blacky got his ride. There was no way he could have known I would not drive all the way back to the house and put him in it again. How could he know I was running late? I looked straight into his eyes, and he stared right back at me, through me it seemed. This didn't exactly calm me. I pondered the possibility of mental telepathy, even X-ray vision, and drove on unsure of the answer. How else could he have known what I had in mind and how to manipulate me?

Shortly after this experience with Blacky, a friend of mine, a breeder of Alaskan malamutes, shared an astounding experience he had had with his male malamute. My friend was watching television, and the dog lay beside his chair. Then my friend decided to eat some cookies. Before he left his chair, the dog bounded into the kitchen. By the time my friend arrived, his pet was waiting in front of the cabinet where the cookies were kept. The owner explained that earlier that same evening he had gone to the kitchen several times for food other than cookies, and each time the dog remained by the chair. How the dog knew enough to respond at the correct moment was one more experience that brought me to search for a greater understanding of this new world of communication that was opening up for me.

I began to suspect that mental telepathy was possible in animals; yet I speculated that overt signals by people might exist and serve as clues to the pets. I finally surrendered all doubt when another incident revealed that this telepathic communication was not due to mere coincidence or signals.

I was sick with the flu and resting in bed for several days. During my second day in bed, I was awakened by the crowing of five ranch roosters directly under my bedroom window. My head throbbed with each ear-splitting call. I tried shooing them away. Finally, enraged, I threw a bucket of water at them through the window screen. The roosters scurried a short distance away but, to my chagrin, shortly returned. Princessa, my German Shepherd, sat watching the scene.

The noise became torture. I took hold of whatever I could throw, and shoes and slippers went flying at the roosters from the doorway. I dragged my weary body back to bed. Princessa still watched. I was barely under the covers when the roosters were at it again. I felt totally defeated and buried my head underneath my pillow, fighting back tears. In a little while, Princessa started to paw at the bedcovers. As she continued to tug, I suddenly became aware of the silence—silence! Curiosity finally overcame me, and I followed the Shepherd's lead out of the house. As we rounded the corner, I was shocked!

Princessa had silenced the roosters permanently. They lay on the ground in a dead heap. I was sick at the sight. And baffled. Princessa had been trained not to harm the livestock. She was punished if she did. I was shaken by the experience. How did Princessa understand? Why did she risk a reprimand? How could she know how ill I was and that it was only those five particular roosters that were causing my pain? We had sixty hens and roosters all milling about, but they were not disturbing me as those five had been. (I had raised them, and that's how I was able to recognize them.) Meanwhile Princessa was prancing about, wagging her tail, kicking up dirt. She was so proud of what she had done! I looked at her and told myself that dogs do not experience feelings like joy and pride at doing something to alleviate a human being's pain or irritation. Or do they? The violence of the act shattered me; yet there was a lesson to be learned here deeper than the loss of five roosters. It was a turning point. I accepted the fact that animals can react to the mental wave lengths of human beings.

My conviction was soon strengthened by reading J. Allen Boone's *Kinship with All Life.* As early as 1954 Boone recognized mental abilities in animals and pointed out an important distinction: What was often passed off as coincidence in animals was actually thought patterns from them that revealed levels of intelligent communication. Boone's book recounts his experiences of learning to communicate with the famous German Shepherd dog, Strongheart, who, as a major movie star, was placed under Boone's care. During the early weeks of his work with Strongheart, Boone tried unsuccessfully to master the dog but discovered Strongheart had a strong will of his own. Only when Boone reversed the time-honored concept of "man-trains-dog" to "dog-trains-man" was he able to get through to Strongheart. Boone's premise did not go unchallenged. There were plenty of skeptics then too, but he held to his conviction.

That behind every object which the senses can identify, whether the object be human, animal, tree, mountain, plant or anything else, and right where that object seems to be, is the mental and spiritual *fact* functioning in all its completeness and perfection. This spiritual fact cannot be recognized with ordinary human eyesight, but it is always apparent to clarified inner vision.

Inner vision, that was it! That's what I was getting in touch with during my stay at the ranch and what was still developing in me. I now realized how much more there is to existence when it is seen through clarity of vision.

"Ask the dumbest beast—he knows that it is so; ask the birds—they will tell you; or let the earth teach you, or the fish of the sea. For the soul of every living thing is in the hand of God, and the breath of all mankind" (Job 12:7–10, LB).

In June 1972, I was invited to lecture on thought communication with animals, at the German Shepherd Club of Fresno, California. Afterward I received a letter from Mrs. Ely Buffin, a member of the club. It was one more confirmation of what was becoming a deep belief and a way of life for me.

Last weekend, some friends and my husband and I took a backpacking trip into the mountains. On our return, I was hiking alone and took a wrong turn at a junction. I had hiked for about two miles when I was confronted by a huge, beautiful gray eagle. He squawked and scolded me so that I stopped to talk to him. He would not even let me speak with his intense scolding but swooped closer and closer. Finally, his apparent concern made me ask, "Am I going the wrong way?" He became totally silent and flew closer, holding himself still and quiet. I felt a need for hurrying; so I ran back down the trail and found my way to the proper path. I had gone two miles out of my way. Had I been just two minutes later, a search party would have gone out after me, and it would have meant many hours gathering up everyone. Your "gift" has given me the knowledge of how knowing, thinking, and feeling animals are, and I wish to thank you for giving me the sense to listen to the warnings of my new found friend, the gray eagle.

Ely Buffin
3615 E. Clinton Ave.
Fresno, California

Mrs. Buffin had the sense to "listen" to the gray eagle and interpret him correctly, thus avoiding a dangerous crisis. She was given a glimpse of another world and knew it. We all possess sensitivity and inner awareness, but she was directly open to it.

When I received her letter, I felt my confidence in my own abilities grow, and I felt my first tentative theories deepen to belief. My aim in writing this book is to help everyone become aware of his or her potential: the knowledge that each of us has the ability to tune into the sensitive world within, become one with nature, and then hear, just as the animals hear—insects as they walk across the tops of leaves, grass as it grows.

2

EARLY ENCOUNTERS

After this first awakening I deliberately sought situations that would offer possibilities for communicating with animals. When my animals were sick and had to go to the veterinarian for treatment, I would arrive early and spend time in the office waiting-room, mentally communicating with the dogs and cats there. In absolute silence I would mentally ask each animal what was wrong. I would then speak to the owner and discover that the animal's pain or trouble was exactly what had been communicated to me.

But during a visit to the veterinarian hospital I grew to understand animals' physical reactions as well as emotional feelings. One day I looked into the eyes of an old, sickly gray-and-white cat; a sudden wave of nausea came over me, and I felt a stabbing pain in my kidney area. I knew there could be nothing wrong with my kidneys because I felt fine, but I allowed the feeling to grow to see what would happen. I felt sure I was vicariously experiencing the cat's physical pain in its kidney as well as a strong feeling of being old and tired. I had the sensation of wanting to die, and as I thought of death, I felt sleepy and peaceful. It began to dawn on me that the cat was telling me it was in great pain and looked forward to dying. There was no fear of death, just a sense of peace. I asked the owners what was wrong with the cat, and this was the reply: "His kidneys are very bad, and we've brought him in to be put to sleep."

The experience is like a dream. In a dream you see, hear,

taste, and feel, but it is all going on inside your head. Using my special awareness I experience, see, hear, taste, and feel what the animal experiences, but it is all going on inside my head. At the same time I am wide awake and aware of everything around me. Someone took a movie of me once while I was working, and the only change they observed was that while I was looking inward to the pictures in my head my eyes went out of focus.

I must admit, however, I was terribly confused when I started receiving mental impressions from animals. These strange sensations and feelings shook my sense of logic and my sense of reality and made me feel isolated from other human beings. In the early days, when I told my friends about my mental communication with animals, most tended to be kind or tolerant, and some were patronizing. As my strange talent became more widely known, I experienced the gamut of reactions from ridicule to skepticism, from laughter to suspicion, from embarrassment to open hostility and unconcealed doubt about my sanity.

Fortunately I learned that Fred Kimble, a psychic who lived in California, also communicated with animals via ESP. I attended one of his lectures, and it was soon clear that he believed the source of his understanding sprang from occult inspiration and was supernatural. I believed mine was *natural.* I did not attend his classes, but I was greatly relieved to know that I was not alone in this ability.

After a period of about six months I began to understand more clearly how the communication worked. Also the accuracy of my nonverbal reception from animals increased. I lost my fear of criticism and began to share what I knew with my friends, most of whom are animal lovers and owners. They were fascinated with what I told them about *their own animals;* in turn, they shared this information with their friends, who soon spread the news. Gradually feature reporters from papers like the *San Gabriel Valley Tribune,* the *Los Angeles Times,* and the *Los Angeles Herald Examiner* wrote about my work with animals. Invitations to speak at dog clubs began to pour in, and eventually I was appearing on radio and television programs. Then appearances on the Mike Douglas television show along with other network shows gave my work national exposure. Newspapers and magazines requested more interviews. My credibility was accepted, and my level of accuracy in consultation with animals anchored itself on solid ground.

Meanwhile more and more people called for appointments with their sick pets. Most of the animals had been taken to veterinarians but were not helped. Possibly this was because many of their problems were emotional instead of physical. Each case was a challenge. I kept refining my technique, and my private practice grew.

One memorable case was Poi, an Australian Shepherd dog. He urinated in the house, ate tissues, chewed furniture, and was covered with a rash. He also suffered from edema, or some form of body swelling. So far all medical treatment had failed. When his owner, Mrs. Griffin, brought him to me, I asked Poi what was wrong. He told me he didn't use his dog door because a bird attacked him and pecked at him whenever he went out into the yard. He also said that his owner was sad and cried a lot, and he wanted to comfort her. He was sad and lonely, and he wanted a companion to play with during the day so that he could romp around. After our conversation I spoke to Mrs. Griffin, and she verified everything. There was an attacking bird in the area; Mrs. Griffin had recently lost her husband; and since then Poi had been left alone a lot. Poi had shown both his own needs and his empathy with his owner: He had literally grieved with her by eating the tissues. The frustration of loneliness, the desire to help his owner, and the undischarged energy were all causing his physical symptoms, excessive itching and swelling. The owner got rid of the bird by moving its nest and got a companion for Poi. In almost no time the rash disappeared, and Poi was going outside to urinate and living a normal, healthy life.

Another memorable case became a personal breakthrough. Since I vicariously experience what the animal experiences, if a dog has a headache, *I* have a headache. I realized for my own health and sanity I needed to close myself off to a degree from some animals' physical and emotional problems. I had to learn to communicate with the animals without doing their suffering. I had to learn to separate myself from their pain, or *close off pain identification.* This posed a real problem, and it was Casey, a Boston Bull Terrier, who dramatized it for me and forced me to discover what I now call my technique of pain-separation, or *closure.*

When Casey was brought to me for help, his owner, Rachel Hill of Santa Ana, California, was receptive to the idea of using ESP. Her only concern was to discover why her normally affectionate dog had suddenly become an unpredictable, uncontrollable animal, barking and biting anyone without warning, even biting off his owner's fingertip during a frenzy.

I mentally asked Casy why he was behaving this way, and he explained that hearing noise of any kind, even the doorbell, was so painful that he lost all control of his behavior. He was terribly ashamed and sorry for biting his owner. I received the image of a brain problem, causing an unbearable pressure that intensified whenever sound reached it. I took this to be the source of Casey's pain. When Mrs. Hill took Casey to his vet, the doctor rejected my diagnosis because he could find no pupil dilation or loss of coordination, some clinical signs of brain damage. Casey's doctor was unable to X-ray the dog's skull because he did not have adequate equipment. An electroencephalogram could have been taken or he could have recommended a doctor with the proper X-ray machine. Unfortunately such was not the case, and within a short time Casey died of the brain problem indicated.

When Casey left my home after that first consultation, I developed a headache of such intensity that I had to take an aspirin and lie down. When the headache still did not diminish, I began to question its source. I soon realized my headache was the same as Casey's symptomatic pain, and I was able to relinquish it. I learned how to close off pain-identification by telling my subconscious that the pain belongs to the animal and not to me. I then ignore the symptoms, acting as though they do not exist, and they immediately disappear.

Soon after this important discovery I received the following letter from Rachel Hill. It gave me much comfort and encouragement. Apparently she had taken Casey to the vet again for further treatment. And now . . .

Our vet confirmed your diagnosis of a brain tumor. Casey began having trouble with his coordination (which I think you noticed), but which we didn't pay too much attention to as he has always been kind of floppy and awkward, but last evening it seemed more pronounced so I took him to the vet this morning. The vet is keeping him for a couple of days, but these are just stop-gap measures, we know. And so tonight our hearts are very heavy and my tears have made my eyes red and sore. But all the tears and all the heartaches are unimportant compared to the pleasure he's given us in the years past. He was a funny little clown with a slap dash sense of humor and above all so devoted to us and so happy for some petting and attention. It was because I knew he loved me that I could never bring myself to do anything about his biting me. I always knew he didn't mean to hurt me and I'm more sure of it now than ever before. Thank you for your help and using your great gift to help all animal lovers.

Rachel Hill
Santa Ana, California

Despite such successes and a growing reputation, however, I am often asked what qualifies me to be an animal analyst. My response is that I cultivate the gift of a free, open spirit that allows animals to speak to and through me. My answer may get an indulgent smile or a quizzical look, but I could not be more sincere. Contrary to popular belief, ESP is not limited to a gifted few but is an innate human quality. I believe everyone possesses the ability to communicate with other forms of life and with each other through ESP. The time has come for this to be freely and responsibly acknowledged.

The basic difficulty in discovering one's ESP power lies in reawakening an ability that has been allowed to become dormant. There are no set rules that must be followed for this reawakening, but different methods are available, and part of the process for each individual is finding what method works best for him or her. There are no economical, social, or educational barriers. Obviously, in my own case, formal education had little to do with awakening my ESP abilities. I hold a bachelor's degree in Bible education from Columbia Bible College, Columbia, South Carolina. It is only *since* my involvement with animal healing that I have undertaken serious scientific studies, including two years of chemistry, physics, zoology, genetics, and other science courses at Citrus College in Covina, California, and at California State Polytechnic University in Pomona, California. Such studies have given me a knowledge of animal anatomy needed to relate animals' physical symptoms of pain to veterinarians. My teachers have also been the animals themselves. It is they who have helped me become increasingly receptive to their messages in this exciting and exacting work. I use the word *exacting* because only through accurate communication can I assist an animal, particularly in cases which require a probe spanning several years or one dating back to the traumas of birth itself.

Immediately after diagnosing Casey's problem, I realized it was essential for me to further my knowledge of anatomy so that I could accurately describe the problem, as I saw it, to the veterinarian. Indicating with my index finger or telling the vet the animal hurts "here" or "there" was hardly a sound or professional approach. So I undertook further education.

Two particular courses (one in organic chemistry and one in anatomy) and two particular dogs (both German Shepherds, one being my own Princessa) showed me how formal study really complemented my ESP abilities.

A German Shepherd was brought to me with one of the more unpleasant and puzzling problems of dogs—eating sticks, stools, and gravel. (Princessa too had been eating stools and gravel.) The owner said she had taken the dog to the vet, but nothing could be found that shed light on this disturbing behavior. I mentally visualized the animal's stomach, and an image of tiny ducts of some sort entering the stomach materialized; they were nearly empty, almost dry. It was so clear I felt like a camera observing a scene, able to zoom in on a particular area. Having no idea what to do with the image, I admitted as much to the owner and was unable to offer any advice.

Later while I was sitting in my organic chemistry class, the solution quite unexpectedly surfaced. As the professor lectured on the digestive process of the human body, he explained that hydrochloric acid (HCL) is secreted by the body and passed through tiny ducts into the stomach. This secretion disintegrates the proteins in order to make these particles small enough to be assimilated by the body as they reach the small intestines. I was seized with excitement! Those were the same ducts I had seen when working with the Shepherd. The dogs' stomachs (both the Shepherd's and Princessa's) were not producing sufficient amounts of HCL to effectively break down the ingested proteins. These fermented in the intestinal tract, producing uncomfortable amounts of gas soon after eating. Since the stomach was unable to do its job, the dogs were eating gravel and stools as an attempt at relief. When I discussed this with our vets, Dr. Darrow and Dr. Christiansen of Arcadia, California, they explained that the dogs were looking for the HCL used in the digestive process of healthy animals which is present in their excretions.

To correct the problem, the vet suggested using salt, which contains chloride; it did not work. He then recommended an enzyme supplement which also proved fruitless because a stool sample analysis had already shown that Princessa's pancreas was producing amounts sufficient to dissolve carbohydrates. I decided that if HCL was lacking, then why not give her precisely what she needed? I purchased some HCL tablets from the health food store and gave Princessa one tablet about ten or fifteen minutes before her meal. Within two weeks, the symptoms disappeared. Another interesting discovery I made while using HCL was that when it was given with dry food with water mixed into it the water diluted the already deficient amount of HCL in the stomach and intensified the problem. But when the food was fed dry with water not available until after feeding, I was able to reduce the amount of HCL supplement needed.

Soon I was working with veterinarians as I do today. We are a sort of team, complementing each other's knowledge. I tell them where the animal hurts and how the pain is conveyed to me. They determine the diagnosis and decide the treatment. Many times a vet knows an animal is in pain, but he does not know where it originates. I think of myself as an interpreter for the animals and an aid to the doctor.

A Boston Bull Terrier named Mickey was brought to me with a unique problem. Mrs. Vivian Wolf of Anaheim, California, brought the dog to my office to see if I could determine why the animal was acting as if he had pain in his nose. The vet could not find any reason for it. Mentally I asked Mickey what had happened, and through a movielike sequence of mental images, the Terrier explained. He had been sniffing around during a walk in the desert when he suddenly felt a sharp stab in his nose. This set off a series of sneezes that did not cease for about forty-five minutes. When they finally did, his nose hurt badly. I vicariously experienced Mickey's pain as an insect sting of some kind; I also realized that the insect had been expelled during the sneezing. By the time Mickey had been brought to the veterinarian, there was nothing visible in his nose to pinpoint the problem. My interpretation unlocked the mystery. Sometime later I met Mickey's vet, Dr. Richard Dahlum of Orange, California, who told me that my information had enabled him to successfully diagnose a similar case.

Mickey also had a chronic ear irritation that was being treated with liquid medication. When I mentally looked into his ear, I saw a form of bacteria burrowing into the moisture. Mickey had allergies that produced excess moisture in the ear canal, with the result that it became a perfect breeding ground for bacteria. Since the medication could not penetrate the moisture to reach the bacteria, I suggested an antihistamine along with the medication to dry out the ear canal so that the medication could reach the bacteria. Dr. Dahlum agreed, and the ear problem was solved.

Sometimes a physical problem may be the source of an emotional problem. Bala had been bred as a cross between an Australian Shepherd and a Dingo dog, a wild dog in Australia. The cross in the breeder's opinion makes a better herding dog. The owners thought Bala was stubborn because he did not respond to their training. When Bala and I communicated, he explained that he wanted to learn but he could not hear their commands. I mentally looked into his ears and saw large wax plugs had built up in the ear canals. I also saw that the dog had no ear drums. The vet suggested removal of the wax by surgery at the Veterinary Teaching Hospital in Davis, California. The operation was performed. When the wax plugs were removed, it was discovered that the dog had no ear drums and no scar tissue to indicate he had ever had any.

Another case that I think of as a real breakthrough in my work concerned a Poodle with erratic behavior. One moment the dog would be lying down calmly, and then, if anyone, including his owners, touched him, he would leap up and attack them viciously. At other times he would suddenly attack the straw of his igloo-shaped basket. No medical reason could be found, and the owners were in despair. When I began to communicate with the dog, he revealed an experience that had occurred *at birth.* I received an image of a newborn puppy lying with his mother under the bed where she had gone to give birth to the litter. I saw her trying to release the puppy from the embryonic sac but failing. I experienced what the puppy had felt as he gasped helplessly for breath, the life-sustaining oxygen still denied him in the sac, and felt hands pulling at him. The Poodle's vet was Dr. Howard Kurtz, who has since become my own trusted vet, and I suggested the owners relate the image to him, along with my explanation, or "diagnosis": possible brain damage due to lack of oxygen because the puppy had been left in the sac for so long. Based on my findings, Dr. Kurtz treated the Poodle for a possible condition of epilepsy, a frequent result of brain damage. The attacks stopped. As a follow-up, the owners called the breeder of the dog and received undisputed confirmation. The breeder said that two puppies were born and that one died before they were able to render assistance. The surviving puppy was taken, by hand, from his embryonic sac only after they were able to extract the mother and puppy from under the bed.

Dr. M. J. Cain of Cincinnati, Ohio, is another veterinarian with whom I now work, after having voluntarily submitted to various tests. Dr. Cain has a farm in Kentucky where he keeps race horses and also treats horses with acupuncture. He had several of his patients there and had already X-rayed their injuries and diagnosed their problems. As a test I agreed to describe their problems through my ESP abilities. I will cite just two examples.

The first was a horse trained as a jumper that had been forced to end his career because he came up lame after a series of jumps. The horse told me he enjoyed jumping, but the pain afterward was so great he was forced to limp. When I mentally looked into his front legs, I saw affected splint bones, and I also felt him landing very heavily on his front legs. He was a large horse, and he did not shift his weight quickly enough from front to hind legs; so the front legs took the shock and weight of the landing, and the splint bones were far too delicate to handle the sudden pressure. Hence the split bones, the pain, and the limp.

Dr. Cain told me he had reached the same conclusion through leg X-ray.

I then worked with Dr. Cain on a dog case in which X-rays did not reveal any specific information. The dog continually chewed at its foot, and nothing appeared to be lodged there. When I mentally entered the dog's foot, I saw a tiny glasslike substance embedded between its toes. Dr. Cain investigated surgically and found an area of scar tissue that could have held such a tiny particle. After Dr. Cain removed the scar tissue, the dog stopped the chewing.

So much for animals in pain or animals with problems! They have helped me help them, and the experience is rewarding to all. They have also, however, enriched my life in a much deeper way. My lesson began (as so many have began in the past and still do) with one of my own animals. My teacher was Princessa . . .

3

ANIMALS ARE ANIMALS

It was one of those rare autumn mornings when light and color are perfectly blended. As I sat on a hillside in the California sunshine studying for a final chemistry exam, Princessa lay near me, content to share this almost lyrical moment. My eyes rested on the back of her head, and I thought: *How bored you must be! I get to do so much, and you have to stay home alone, day after day.*

Slowly Princessa turned her head around, looked me straight in the eye, and communicated a direct answer: *But my life is wonderful! I have time to see the things you are too busy to see!*

What do you mean, Princessa?

Hear that bird? Listen to that dog! There goes that car again, coming our way. Just smell this earth! Hey, look at that rabbit! There goes another bird!

These sights and sounds were out of my immediate physical range. But once I took Princessa's advice and began mentally listening through her ears and seeing through her eyes, the countryside became an enchanted ground, filled with exciting life. Princessa could not possibly be bored! As we sat there on the hill, she told me about her day. It was literally filled with activity and play, sights and sounds and sensations, and excitement. Sometimes she ran about, retrieving sticks the ranch hands threw for her. The men tired quickly, but she flew over the ground, stretching her muscles with the sheer joy of racing and running. When the game was over, she returned to the comfort of her own world, savoring an old bone, smelling the earth, experiencing the subtlest puff of wind.

Since this first lesson, I have learned to look at the world through animal eyes, far beyond the range of human sight and hearing. The secret is to stop long enough to absorb what is going on around me, even in the *desert.* I was in Las Vegas once, playing bingo, and I asked a lady from the East Coast if she had visited Death Valley yet. She looked at me with horror and disgust. "Death Valley? It's nothing but sand. How boring. And dull. Las Vegas is where I belong!" It takes all kinds, as the saying goes, and this lady and I are opposites. I find Las Vegas a dull, unchanging place, though plenty of money changes hands it leaves the soul empty as well as the pocketbook. But when you travel through the desert, Death Valley in particular, and take the time to really observe what is there, you'll find it is always changing. The beauty is endless, and the serenity uplifts the soul. I spent a week there recently, sitting for a couple of hours every day just listening and looking through my animal-trained eyes. The desert is teeming with life if you look beyond the heat and sand.

Yet until Princessa's lesson I was just as rushed as the next person, so intent on tomorrow that I missed the experience of today. Now life is exciting, and the most everyday experiences have a dimension that were totally lacking before. Now living "only for the moment" is a cliché for me because I have been given the grace to experience what it means to live "within the moment."

In real life and in literature most people are quick to interpret animal behavior in human terms. Princessa taught me not to do this. Animals have different motives; we should let them be. Animals are animals, and they are much more exciting as animals than as actors in Aesop's fables or the *Cat in the Hat.*

The biggest mistake I made before learning to communicate with animals was believing that being human was a far superior state of existence. This idea is shared by many animal lovers who say and really believe, "My dog thinks he's people." It's not the dog who thinks he's people; it's people who give humanlike qualities to the dog. When I asked animals if they wanted to be like people, they were emphatic in telling me they were perfectly content to be what they are. Their behavior resembles that of humans at times because it is their way of making their wants and needs known and their attitudes understood.

Although I now know how stupid it is to consider an animal in human terms, I have been just as guilty as the next person. After all I had learned from Princessa, I reverted to this error. I was trying to teach Princessa to speak by training her to bark for something she wanted. I was enthusiastically saying: "Come on, Princessa, speak! Say woof, woof!" I kept repeating this so-called training instruction until I was hoarse, but all she did was sit there with a puzzled expression, her head cocked to one side, refusing to utter a sound. Finally in exasperation I mentally asked her: *Princess, what's the matter with you? Why don't you speak?*

She mentally answered: *You already know what I want. Why do I have to bark?* Her answer was so sound and so sensible, I was ashamed that I had tried to reduce her to human "doggie" talk.

It is vital to accept the differences and similarities between animals and people in order to pave the way for communication. Accepting the animal as animal and valuing it for that alone is the first avenue to receptive listening and response. While some information in this chapter may appear elemental, it may come as a surprise to discover how many intelligent and well-meaning pet owners disregard the obvious.

My anatomy studies taught me what similarities there are in bodily functions between animals and humans. Both species require sound basic nutrition, and both reproduce life in the same miraculous way. Animals fall prey to many human diseases, and research experiments on them are often responsible for discovering cures that benefit humans. Experiencing emotions is another bond between animals and humans. Animals have told me they are capable of feeling love, hate, resentment, frustration, jealousy, depression, and joy. But one attribute that sharply differentiates the two is conscience. Humans wrestle with the question of morality. It is different for animals. While a pet may feel upset after he has done something wrong, his guilt does not stem from a sense of morality but from the reprimand. This sense of conscience in man I call spirit, or the thinking process, which includes an ability to make intricate plans for the future.

Animals cannot do this. Ah, I hear you saying now, "But they do plan for the future. Look how animals store food for the winter." That is instinct and involves no reasoning or special mental process. I even question how much of an animal's behavior is actually instinct and how much is taught by the mother. Behavioral psychologists are just beginning to discover that some animal behavior, once branded as instinctive, is actually taught by the parent. (This subject is covered at length in chapter 8.) Animals are capable of sensing a situation and acting accordingly, but they cannot, as humans can, decide where they want to go on a vacation and execute all the necessary preparations. An animal is not capable of such complicated thinking or action.

Another important area of difference is environment. Environment shapes human lives to a large degree, but our growth does not solely depend on it. When we realize we have a problem resulting from environment, we can decide to do something about it and, by various actions, bring about the change. Animals are under the control of their environment. Even though they sense something is wrong or unhealthy or dangerous in their environment, they cannot make any changes without the help of people. An animal can escape a situation by running away if the opportunity presents itself, but it cannot alter an environmental problem the way a human can.

Experiencing time is another difference between people and

animals. Animals are not aware of time in the way humans are. During our communications, animals will tell me about an event that occurred one year ago and then jump to something that happened yesterday. They do not experience time sequentially. However, if I ask specific questions about a particular event, an animal can give me the information in chronological sequence.

Awareness of time duration is another difference between humans and animals. An owner can be gone for ten minutes or an hour and still receive the same greeting from a pet upon return. (It "feels" the same.) I am speaking here of the emotional aspect, not the physical. Confinement over a long period of time will create a build-up of undischarged energy in an animal which may produce physical symptoms of frustration, but this reaction is not related to the animal's concept of time. However, absence from an owner for an *extended* period of time, such as several weeks, will produce a homesick longing in a pet. This is because it is aware of missing someone loved, not because it is aware of dates or days of the week.

Horse breeders frequently want me to ask a horse about a show on a particular date. This is impossible. A horse has no idea what the "fifteenth of the month" means. But if the owner gives me a specific experience which identifies a particular show, I can ask the animal about an event. For example, a question such as why the horse refused to jump at the third fence can bring a specific answer.

A common misconception among animal owners is that an animal possesses a sense of time because it does the same thing every day at the same time—such as waiting at a window at 3 P.M. for the children to return from school, or going to the train station daily to wait for an owner. There are two explanations: One, emotional; the other, physical. Emotionally, the dog is "tuned into" the children and knows they are on their way home. In the case of the dog waiting at the station, there exists a biological "feel" about that time of day. This same physical or biological principle holds true for cows at milking time. The cow feels the udder fill with milk and "knows" that it is time to be emptied.

A great difference between humans and animals concerns the question of sex and sexuality and, of course, propagation. There is considerable controversy and confusion over the question of spaying and neutering animals (that is, birth control), along with the pros and cons of allowing one's pet to breed because it is believed to make a better adjusted animal when she has had one litter and then is spayed. This is an old wives' tale. I deal with these subjects in detail in chapter 10. Here I want to report briefly on the biological and psychological differences between animal and human sexuality and reproduction because, according to my animal friends, we humans know very little about what *they* feel about sex and propagation.

From what the animals tell me, males and females alike, sex is a tremendous *physical* pressure, and there is absolutely no emotional content to the experience. Male animals, for instance, do not seek females for mating unless the female is in heat. Sex then becomes a strong, uncontrollable physical drive, like hunger for food, and not an act of love or enjoyment. Spaying and neutering automatically remove that pressure, thus freeing them to be more loving pets.

The same applies to propagation. My communications with dogs and cats reveal that 99 percent of them *do not want litters.* This is no exaggeration. Physical drive is something animals cannot control, nor do they connect it with the end result of puppies and kittens. Many humans, on the other hand, tend to identify their sexuality with that of animals and feel that because sex is important and enjoyable for them it must be important to their pets.

Animal mounting is another commonly misunderstood behavior. When two dogs meet, and one animal mounts the other, it is an expression of dominance. There is no sexual connotation to it at all. So if two males do it, there are no "homosexual" overtones. When one dog successfully rides another, then the "rider" becomes dominant in their relationship. I have often seen this when my spayed bitch and her neutered son play together. If the son has a toy she wants, she will ride him as a way of telling him to surrender the toy. This also happens when I am busy and one of my dogs wants attention. If I continue to ignore him, he will try to "ride" me, forcing his way of "domination" on me. Whenever I have observed this "mounting" or "riding" and asked the animals about it, they have revealed its nature clearly. It is a matter of domination, nothing else.

There are exceptions, of course, and these should be checked. The male dog who habitually "rides" a person may have an overactive hormone stemming from an enlarged prostate gland. He cannot control his behavior and should be examined by a veterinarian. According to veterinarians with whom I have discussed this problem, the only solution is to have the dog neutered as soon as possible. The younger the dog is neutered, the better for the animal. In older dogs surgery can be fatal. Prostate problems and prostate cancer are common killers of older male dogs.

Sniffing at people, particularly in the genital area, has no sexual connotation. An animal identifies a person by body chemistry, which is unique to each person. In the same way animals also identify one another and learn who has been in the neighborhood.

Summing up, sex for the animal usually means reproduction, like it or not, and this is one good reason to understand that our pets are animals with their own needs which are distinct from human needs.

It is wonderful to communicate with animals. It is a mistake to confuse them with humans because they tell me they truly enjoy being themselves and have no desire to be like us.

4

NONVERBAL LANGUAGE

As I have explained earlier, ESP, or nonverbal communication, is a natural ability we all have as children. It's something we're born with, but it is lost later in life as we learn words and focus on the use of words alone for communication. A visit to a special education school in Northern California in July 1974 provided an excellent example of this natural ability. One student was a twelve-year-old boy who had cerebral palsy and had never been able to verbalize anything beyond grunting noises. We met, and as I mentally received his thoughts, he ran away as fast as he could.

"I can't understand that," one staff member said. "He always greets strangers with affection. Sometimes he's so demonstrative he makes a pest of himself." The staff member tried to coax the boy out of the corner of the room, but he refused to budge. I stayed where I was, but across the distance I mentally asked the boy, *Why did you run away from me?*

I was not surprised when he replied in the same, silent communication. *I can read their minds,* he answered, referring to the staff. *That's how I get my way. They want me to work, but I like just sitting still. Now somehow you know my secret. I was afraid you were going to tell on me. Then I wouldn't be able to do it any more.*

I talked with the staff members, and gradually the youngster relaxed and left his corner to join us. Wanting to test the validity of this communication, the staff members agreed to an experiment. They all mentally told the boy to mop the floor. The boy had already admitted to me that he was lazy and didn't want to work. But he read their thoughts, and his eyes widened and he grunted, "Uh! Uh!" He shook his head and ran away again.

You may think this child a special case, but your experience will show you many examples of normal children communicating nonverbally among themselves, with animals, and with adults. Haven't you seen a small child look at an adult and immediately love or dislike that person for no apparent reason? Some adults who love children elicit an immediate happy response from children without doing anything special to entertain or win them. The child is using his or her natural ESP to look into the adult and discover what the adult's *true* feelings are. You may also see children from two different countries (parents speaking different languages) play together for hours with no communication barrier. They are basically conversing with their minds. Unfortunately, once children begin to verbalize feelings, this marvelous ability ceases to be used, becomes dormant, and by adolescence may be lost entirely.

Young children who still use this nonverbal language can also communicate with animals. Two families, both friends of mine, illustrate this. Mr. and Mrs. Moen of Lompoc, California, and their young son, Trajen, are very good friends of Mr. and Mrs. Stan Appelt of Bradbury, California, the owners of Burgermeister, a black German Shepherd. Trajen and Burgie had a beautiful ESP friendship for the first four years of Trajen's life. The little boy often came into the house to tell his mother, "Mommy, Burgie says he's thirsty," or "Burgie hurt his leg, right here," pointing to the area on his own leg which Burgie had conveyed to him was painful on *his* leg.

This sounds like childhood fantasy, but when it was checked out, we knew it was genuine communication: Burgie was indeed thirsty or hurt in some specific way. Then the friendship was interrupted when Trajen's father took a year's business transfer to Florida. During that year Trajen entered school and developed a much larger vocabulary. Trajen and Burgie were reunited when his father returned to California and visited with the Appelts, but something had happened to their ability to communicate. After twenty minutes with the dog, Trajen ran into the house in tears. "Mommy, why doesn't Burgie talk to me any more?" When I visited with the boy's parents at the Appelts' home, I sensed Trajen's hurt and dissappointment and Burgie's deep sorrow; and I explained to the parents and the dog that Trajen had lost his ability to communicate nonverbally. As the little boy outgrew nonverbal language, a special friendship ended. Unfortunately, at that time, I did not know how to help the boy recapture his ability.

ESP communication with animals is not difficult to achieve but is deceptive in its simplicity. To the degree that one believes this language works, it will work. How much you are able to open yourself to an animal determines how effective you will be. If you have reservations, communication through mental pictures will be less than accurate, and frustration will result. The secret of ESP is to let go—it's that easy and that complicated—just let go of self.

The key to understanding this language is to understand the use of visualization, or mental pictures. This is crucial. When we speak or think, we always have a picture in our minds. For example, when we speak of going to the lake, we have a picture of a lake in our minds; it is not necessarily a specific lake but a lake of some sort. This picture is then transmitted by ESP to an equally sensitive mind of the person who still has this ability—the small child, the nonverbal person, or the animal. We also experience joy, pleasure, or disgust about the lake. This is transmitted. For example, when my animals, who normally love people, back away from somebody, I never force that particular confrontation. The person may be saying to me, verbally out loud, "Oh, I adore animals! Especially German Shepherds!" At the same time he or she, while trying to impress or please me, is thinking, *I cannot stand those big dogs.* My animals are not hearing the verbal words; they are mentally seeing the nonverbal message in terms of pictures. The person is thinking, *I hope that nasty, filthy dog stays away.* The disgust is instantly transmitted to the dog, and it gets a picture of itself backing away. The person is using ESP without even knowing it! He or she is communicating through visualization, or the transmission of mental pictures, an innate ability that has lain dormant for many years, depending on the individual, but can be reawakened, learned, and used. If you want to communicate to your pet, here's how.

Get your favorite camera. Take a picture of your pet in the position you want it to obey. For example, if you want to teach your dog the "Down! Stay!" position, photograph it while it is in that position. Memorize that picture until the image is clear in your mind. Then put it away. With practice it will become easy to create that mental picture at will. It is like taking a still photograph with your mind—snap it and hold it. The next step is to visualize this picture while giving your dog the command. For example, to train it to understand the word *sit,* tell it to sit while mentally visualizing it in that position. Soon the animal will respond to the command as it receives the image from you. While this instruction is going on, ESP communication is taking place. You are conversing in the animal's language; as you become more proficient, chatting with your pet will seem natural.

There are a few rules but not many. First, always call your animal by name to be absolutely sure you have its attention before you start visualizing and giving the command. Second, verbalize the command. This makes the command clearer *in your own mind* because we naturally picture what we're saying. Thinking *without* visualization is nearly impossible. Last, avoid using the negative. Animals tell me they don't understand the negative. For example, words such as *don't, isn't, can't, not, shouldn't,* and *wouldn't* are abstract thoughts which cannot be visualized and therefore cannot be understood. So when you tell your dog, "Don't get on the couch," the dog throws out the word *don't* which it cannot understand and acts on what it does understand, which is "Get on the couch." While you are saying "Don't get on the couch," you are actually visualizing the dog getting on the couch. The dog receives your picture of it's getting on the couch and obeys. You wonder why. In a few rare cases the dog will obey the negative command, but this is because the owner who is saying "Don't get on the couch" is actually *visualizing the dog on the floor.* So be positive. Tell your pet what to do instead of what not to do. Instead of "Don't get on the couch," say, "Stay on the floor," visualizing it there. You may also say, "If you get on the couch, I will punish you," while visualizing the dog jumping onto the couch and then getting spanked or put outside.

It is the same with other animals. If you're trying to teach a horse to go over a jump, mentally visualize the horse taking the jump with complete ease. I have given riding lessons based on this thought-communication pattern with successful results.

My first experience in testing visualization was with my Pomeranian, Blacky. I wanted to experiment with mental commands to see how far Blacky would follow. I knew he loved liver-flavored dog biscuits; so while he was lying under the bed, I visualized Blacky coming out and getting a dog biscuit I held out for him. He came. Then I visualized Blacky asking me for another by first going to the box and sniffing and then returning to me and sitting patiently until I handed him another. He did just that. Then I actually gave him the whole box of dog biscuits. An astonished Blacky picked it up and scurried under the bed, dragging his treasure. After I listened to him munch happily for a few minutes, I mentally pictured Blacky pushing the box out from under the bed with his nose, myself looking at the box, then Blacky taking it back under the bed to eat the whole thing. He would not budge but went on hurridly munching. Mentally I said, *If you keep the box, I'm going to reach under the bed and take it.* I visualized the action, slowly, step by step. In a couple of seconds, Blacky pushed the box out from underneath the bed with his nose. I must admit I was somewhat overwhelmed that he had responded so obediently to my mental pictures. He was actually reading my mind. I kept my part of the bargain and gave Blacky the whole box. He was like a kid let loose in a candy store! After this successful experiment, I started using mental pictures as a regular part of my work.

People who show dogs often have difficulties in the ring because they are not aware that their thoughts are being picked up by their animals. For example, when a dog is taken to show for obedience trials, the animal is told, "Down, stay." The handler stands at a distance and waits for that eternal three minutes to pass. He or she verbally says to the dog, "Down, stay," but while standing back, he or she is thinking, *Oh boy, last time we were here he got up and ran out of the ring. I hope he doesn't run out again this time.* The dog picks up the owner's or handler's mental picture of running out of the ring, and it does exactly that. The handler then usually punishes the dog for disobedience. The dog cannot understand this because it did what it was told mentally: *Get up and run out again this time.*

Two common problems can block receptive communication. First, when communicating with an animal, the mind must be kept openly receptive. Never try to figure out what the animal is trying to say; just let your mind go blank. When you are trying to figure out what the animal wants, your mind is so busy that the animal's thoughts cannot get through. Second, it is necessary to block out all other thoughts. The way to accomplish this is to identify with the animal by putting yourself in its place. It often helps to start by looking into the animal's eyes because you can then see and feel its emotions. When you gain experience, this will no longer be necessary. Stop thinking about yourself, and do not try to interpret the animal's actions according to what you would feel if you were in its place. You will gradually feel your presence lessen. Only the dog will be in your mental picture.

For those who are able to communicate, receiving the reply

is often scary at first. It is important to know how the reply occurs. While you stare into the animal's eyes, you will gradually get a "hunch" or an intuition that the animal wants something. You will also experience a change in your emotions. And you may think, at first, that it is your imagination. You will feel silly too, but respond to the hunch. No matter how silly you feel, you will gain experience and will eventually be able to distinguish the difference between your imagination and what the animal is saying. I ask people not to tell me anything about their animals before I talk to them because it is hard to distinguish between my deductive thoughts and what the animal is telling me.

When you wish to ask a specific question, such as why the horse refused to jump a certain hedge-shaped obstacle at the horse show, you should form a picture in your mind of the horse in an English jump saddle with the rider on it, going around a jump course. The picture in your mind is a vague course, nothing specific; it is just set in a show situation like a ring with a crowd watching. Then visualize the hedge fence with the horse approaching the jump and suddenly stopping. As you do this, the horse will receive the picture and send you back the *corrected* pictures; and the series of pictures in your mind will change, taking on specifics, like the hedge shape, and the surroundings of the hedge. You the rider will also feel an emotional change because the horse will be sending you its own feelings at the time it stopped and refused to jump (for example, from relaxed or happy as you were to tension and fear of falling). The use of pictures and the transmission of emotional feelings are always used by younger animals. I love working with horses because I find them wonderfully receptive and communicative.

Another area of reception is purely physical. If I want to know where the animal is hurting and how the pain feels, I concentrate on the area in question. Maybe the leg, for example. Since the human anatomy is nearly the same as other mammals, with similar organs and basic body structure, I become aware of my leg and then become aware of any changes of feelings there. I report these feelings to the veterinarian or the owner, who will render the proper physical aid. This physical aspect of ESP communication really requires a knowledge of anatomy and diseases and should be handled by your veterinarian.

Perspective plays an important role in understanding the mental picture the animal gives in response. To a small animal, a person may appear huge, but to a Great Dane, the person may just be average size. Similarly a hill can look like a mountain to a Yorkshire Terrier, but to a Saint Bernard it is just a hill.

Mental pictures received from animals reflect their way of seeing. Taste may also be experienced through mental communication. I have often asked an animal how it tastes food and what it prefers in its diet. Then I'll tell the owner, "Your dog would like X-brand food because that's what I taste." Owners are shocked because they assume I have actually eaten dog food. Once in a moment of curiosity, I was so impressed with the delicious taste I mentally received from milkbone dog cookies that I had to try one. Just one! I stopped right there. The taste was terrible, nothing like what I had experienced through the animal's mental picture.

Animals also communicate with one another. I learned more about this one day when riding a friend's horse. Dawn, a Pinto mare, asked if we might visit her good friend, Kelly, a horse with whom she had shared a stall when she was injured.

Dawn's owner, Buz Olmstead, of Stallion Springs, Tehachapi, California, verified the objective facts. We decided to find out what the horses had to say to each other. When we approached Kelly, I experienced a sense of greeting pass between them. Then Kelly asked Dawn where she had been, and she responded by showing Kelly the exact places . . . trails, hills, forest. Kelly received these pictures as he stood staring into her eyes. It was like watching a silent movie—incredible!

Kelly told Dawn how lonely he felt and how he wished he could have been with her by showing her pictures of himself standing alone in the pasture and projecting the feeling of loneliness. I experienced two horses exchanging feelings of parting, after which Dawn willingly returned home.

Intelligence is another factor in nonverbal language, and there are several theories about the relative intelligence of various species of animals. Personally, I do not find any one species any more or less intelligent than any other although many people, including members of the scientific community, base the species' intelligence or IQ on its relationship and similarity to humankind. For example, monkeys are considered more intelligent than other animals because their anatomy is similar to humans and they can mimic us in many ways. But mimicry is not intelligence. In my opinion, if dogs had the same anatomy as humans instead of four paws, and horses had arms and legs instead of hooves, they could perform just as well as monkeys. Another way some people measure intelligence is based on the animal's degree of dependence or independence. Cats and horses are more independent and usually relate better to one another than they do to people. Dogs, however, are more dependent on humans both emotionally and physically. People therefore tend to think dogs are smarter than cats and horses. This is both illogical and unfair. Each species should be considered and accepted as equal, and each individial within that species should be judged solely by comparison to its own kind.

There are many different levels of intelligence *within* each species, and these can be measured. My personal criteria of an animal's intelligence are as follows: ability to communicate; depth perception of the world around it; and ability to distinguish and communicate the events of its life in the actual sequence in which those events occurred. An animal of average or above-average intelligence can relate details in sequence; a retarded animal can only relate fragments of events.

For example, a Labrador was brought to me because he refused to obey commands, even nipping at his owner at times. The Lab communicated that he really was trying to learn, but he became confused because he could not remember the meaning of the command for very long. When punished, he snapped out in sheer frustration because he did not know what else to do.

Retardation is a basic cause of an animal's inability to develop love for and loyalty to its owner. A retarded animal may outwardly express its joy of loving, but what it is really expressing is its joy in *receiving* love. Retarded animals are totally self-centered and are incapable of returning love.

Normal animals, in general, operate on love, but they also respond to fear, anger, and distrust. The mail carrier who genuinely loves animals and is not afraid of them sees them in terms of friendship and pleasure, and the dogs he or she meets will be friendly or simply leave him or her alone. The mail carrier who is afraid of dogs will be in danger of attack because the animal will feel that fear, vicariously, and act accordingly. My animals tell me a human emits no scent when he or she is frightened. The animal picks up the person's mental picture of *expecting* to be bitten. The animal may then bite, but it is reacting to the fear in the individual rather than to the individual's scent.

One day my friend Jean Salmon of Covina, California, was taking her German Shepherd to a dog show. Her children and the dog got into the car first. Jean followed, but not without some emotion; she was angry at one of the children and intended punishment. When she reached the car, the Shepherd, who had his back to her, wheeled around and grabbed her arm and clamped down with his teeth. Just as suddenly the dog realized who she was and released her arm and hung his head in shame. Later when I talked to the Shepherd, he explained his action: He had jumped into the car after the children with his back toward Jean; he was unaware of who was coming but acutely aware of the anger directed toward one of the children. So he turned and bit Jean, trying to protect the child. Animals react strongly to human feelings.

Another friend of mine was trail riding on her motorcycle.

Her boyfriend was with her, riding his own bike. As he rode down the wooded path, a German Shepherd darted out, growling and snarling at him. He feared the animal, and the dog reacted strongly to that. A few seconds later, my friend rode by that same spot and once again the dog started attacking. She immediately shifted the bike into second gear and continued riding, thinking friendly thoughts and telling the animal she thought he was a beautiful Shepherd. The dog stopped running after her and sat at the side of the path with his head cocked to one side. No one could convince my friend that the situation was coincidence. She knew the dog picked up her positive, loving vibes.

My work with animals has shown me that you just can't fool them. One day Princessa sat in the back of the station wagon as we drove past a school yard, and she begged me to let her play with the children. I stopped and a ten-year-old girl came to the car. She asked if she could pet Princessa. For the first time in five years, and the only time in her life, Princessa growled at a child. I asked the child, "Are you afraid of the dog?" She answered yes. I explained that Princessa loved young girls, and when she let this idea change her attitude toward the dog, Princessa licked her hand. After a while, they were able to play together.

So far, I have been talking about short-distance communication in which the person and the animal are in sight of each other. However, I have also learned that ESP nonverbal language is not limited by space or distance and can be nearly as effective over thousands of miles. The next chapter explains this phenomenon and how it has enabled me to locate lost animals, reuniting them with their owners.

5

LONG-DISTANCE ESP

When you communicate with your pet using ESP, distance does not hinder effective interpretation. A pet can be in another room or thousands of miles away from its owner, but ESP communication is not affected.

J. Allen Boone, author of *Kinship with All Life,* got an inkling of this phenomenon when he realized that Strongheart always seemed to anticipate Boone's arrivals. His book tells how a friend took care of Strongheart while Boone was away, and, through the friend's reports, it was clear that Strongheart walked to the window and sat there and waited just minutes before Boone's arrival. No outward warning was given the dog, and the time of arrival varied, but Strongheart was constant in his vigils and accurate in every instance.

A client of mine who "dogsat" for a Poodle during the owner's out-of-town trips recognized this same sort of behavior in the dog just before the owner's arrival home. The Poodle would gather up his toys, put them in his box, and then sit by the window and wait. A check with the owner's time of departure (after an absence of several days) confirmed the fact that the dog's behavior coincided with the exact moment the owner had boarded a plane for home.

When I saw my own dogs acting the same way, I had no doubts left about the power of ESP over distances. The first example of note occurred during a visit with my sister Flo, in Bradenton, Florida. I was out for the day and left the dogs at home with Flo. When I returned that evening, she laughed and said, "I know exactly when you started home. You began to leave once, and then you changed your mind and went somewhere else."

"How on earth did you know?" I asked.

She said that at 8 P.M. my four dogs got up from the kitchen floor and ran to the front door to greet me. About twenty minutes later they returned to the kitchen. Just ten minutes after that, Flo said, they went to the front door again. "And then you came home!"

I had indeed started for the house at 8 P.M., but about twenty minutes later I felt hungry and decided to take a side trip to a hamburger stand. I picked up an order and ate it in the camper on the way home. My dogs had been mentally watching me and knew when to expect me.

After this experience I began checking the behavior patterns of my dogs with persons who stayed with them during my absences. I requested accurate records be kept of any change in behavior or eating patterns, and all variations were logged by the day, hour, and minute.

Blacky, my Pomeranian, is a good example. I would leave Blacky with my friend Evelyn Appelt, and though my trips were of varying duration and she never knew when I would return, she learned my travel plans from Blacky himself. Occasionally she would cancel her own plans in order to be home at the time I would want to pick Blacky up. She said, "While you're gone, Blacky sleeps behind the couch and hides. He only comes out to eat. But when you're on the way home, he comes out and begins to play." A quick check proved that Blacky's change in behavior occurred the day and hour I started for home.

Then something occurred that made me realize there was more to this phenemonon: My animals were receiving my thoughts the *whole time I was away.*

I had to go to San Francisco for business and asked Jerry

DeMent, an old friend who knew my dogs well, to stay at my house and take care of them. Throughout the three-day trip I felt alone, lost, and inexplicably frightened. I missed my animals and wished they were with me. The last day in San Francisco, I had to walk along Market Street at 6 A.M. It was raining, and I felt miserable and nervous as I waited for the bus to take me to the airport. In fact, this feeling of aloneness had been with me during the entire trip, and I didn't relax until I was on the plane and safely headed for home. When I arrived, Jerry said he gathered I was on my way because of the dogs' radical change in behavior. During my absence they had been restless and had refused to eat. This was unusual because they had always had fun with Jerry on past occasions and had always eaten heartily. At seven that morning, however, they suddenly started eating again, and the four dogs appeared relaxed. They had actually experienced my tension and relaxed only when I relaxed.

I learned that animals can mentally "see" and "hear" what is going on around their owners while the owners are away. I questioned a number of animals, and they all told me that communication across a distance is exactly the same as communication in the same room.

To test this I visited a number of boarding kennels in order to ask the animals what they were perceiving about their absent owners and how they felt about being left. From what the animals told me, I believe cats can accept the absence of their owners more readily than dogs because cats seem to be more independent. Dogs tell me they have a difficult time being on their own.

While working in Florida in January 1975, I was asked to communicate with two Boxers. The owners had gone to England for a month and on their return were shocked to find that each dog had lost about fifteen pounds. The owners blamed the kennel staff for not feeding the animals, but when I asked the Boxers about it, I got a different story. They told me the food was there, but they sensed their owners' concern over the food and felt something was wrong with it, and so they refused to eat it. The owners then admitted that during their entire trip they had worried that their dogs would not eat. The dogs had "picked up" this worry and acted accordingly.

There is a lesson to be learned from these Boxers. First, if you have to leave your pet, be sure it is in a good, safe place. Then, if you tend to worry, send the pet thoughts of love and reassurance of your return. Then push your pets out of your thoughts and enjoy yourself. They will too.

Most animals can get along well during a limited separation from their owners. If they know you are coming back, they can handle the break. Only if the animals are confined for several weeks *without sufficient exercise* will there be trouble. The build-up of undischarged energy will cause a physical reaction in the pet such as nervous scratching or chewing.

While you are away, animals can also "see" through your eyes; they can "see" your surroundings through this same ESP communication. A client brought his friends' dog to me to see if the dog realized where his owners were and how he felt about staying with the friend in the owners' absence. The dog gave me mental pictures of a service station and a restaurant in a forest of huge redwood trees. We made a note of the day and time of these communications. When the owners returned from the redwood forests of Northern California, they checked their itinerary. They were indeed walking into a restaurant for dinner that not only fitted the description the dog had conveyed but also the time of day.

In the cases described so far, my communications with the animal had occurred when I was actually with the animal. Now I wondered if I too could communicate with an animal over a distance, and so I decided to try. I was giving a lecture at the Madonna Inn in San Luis Obispo, California, when Dr. Werbel, a physician who was in the audience, asked if I could "pick up" or "tune into" his son's Samoyed dog, some distance away. I visualized the dog and mentally called him by name. Suddenly I realized I was "seeing" the dog's surroundings. I recognized the area as being near Lake Tahoe, some four hundred miles from where we were. The doctor confirmed the fact: His son lived eleven miles south of Lake Tahoe. Then the dog "showed" me himself playing with a little blond-haired boy. Next I saw a man lying in bed, obviously in pain, and eating ice. From the mental pictures the dog conveyed to me, I concluded the man was Dr. Werbel's son and that he had the flu. Dr. Werbel shook his head. There could be no "little blond-haired boy" in his son's life since he was a bachelor; and his son couldn't have the flu or he would have called him. Later that evening the doctor's curiosity got the better of him. He called Lake Tahoe. His son admitted to having the flu and had been eating ice to settle his stomach; he also said that his dog often played with the landlord's son, a blond-haired little boy who lived downstairs.

After this and several other successful communications I found that distance was no barrier, and I began consulting long distance with animals and relaying their symptoms. The process was the same as if I were with the animal, but one condition was necessary: I had to know where the animal was located. If I cannot visualize the animal in its proper geographical location, I get a blank mental response. When I'm directed to the right area, the mental picture comes in clear and strong. When I do not know where the animal is located, I send my mind in a mental circle, similar to the circling of a physical radar beam. When I feel a strong tugging sensation, I know I have located the correct direction. I then move in that direction until I locate the person or animal.

This ESP "radar" can draw pets to their owners over great distances. One such example can be found in Alexander Key's book *The Strange White Dove.* The author tells about a family who lived in Florida and then moved to Temple City, California, a suburb of Los Angeles. They had left their cat with a neighbor, but two years later their cat turned up on their doorstep in Temple City. In another case a cat had been left behind with a friend of mine. It disappeared and eventually found its way from Beverly Hills to Ventura, California, a distance of about one hundred miles.

I worked hard, perfecting this ESP "radar" ability, until I felt competent to consult over the telephone.

One of my first phone consultations was with a Hackney Pony, Midnight Ace, a champion owned by D. L. Arkenau in Kentucky. The pony was having trouble with his tail. Three veterinarians had examined the pony, but without diagnosis or success. Arkenau then decided to dock the pony's tail to relieve his suffering—an action which would also have ended the Hackney's show career. Pat Wood, a horsewoman in Hamilton, Ohio, had recently read about me in a magazine and decided to contact her friend, Arkenau, and suggest my ESP diagnosis procedure as a possible alternative to docking. She then telephoned me in San Francisco. I "tuned in" to the sick pony, who communicated what had happened: While he was scratching his rump on the stall door, he felt a sudden prick, like a sliver of wood entering his tail. A pus pocket had formed against the spinal column in the tail; it was deeply embedded and invisible to the naked eye. Mrs. Wood and Arkenau decided to risk lancing the supposed abcess area. When Midnight's tail was lanced, the infection came out and, with it, the splinter.

A year later Mrs. Wood contacted me again, this time about an American Saddlebred she owned which had become partially paralyzed in her back legs. So far her veterinarian had found no physical cause. While Mrs. Wood and I were on the telephone, I mentally went over the animal's body. A scar and an injury at the high point of the rump became "visible." The horse transmitted to me the picture of a heavy board falling on her back, hitting the spinal column between the vertebrae. Nerve damage seemed possible. Mrs. Wood took the horse to Dr. M. J. Cain, the Kentucky veterinarian who uses acupuncture. Dr. Cain found the scar and the spinal problem causing the paralysis; he used acupuncture to treat her, and eventually she regained a certain amount of feeling in her legs. But according to Dr. Cain a complete cure was not possible because the horse refused to roll, an action necessary to make vertebral adjustments on the spinal column. American Saddlebreds are trained not to roll because owners do not want the tail-set broken. The *tail-set* is a surgical procedure in which muscles in the tail are cut and a device is used to set the tail. The tail is then wrapped into a ball and tied in such a way as to hold it in a vertical position.

Since Midnight needed to roll to effect a cure, I went to Ohio to meet Mrs. Wood, with the hope that the two of us, using visualization techniques, would help the horse break her training block. We took Midnight into an arena where there was plenty of sawdust. For an hour or more we visualized what we wanted: Midnight rolling in the sawdust. Although it was natural for her to want to roll, the training block was so strong that every time she seemed about to roll she hesitated and resumed a standing position. After the visualization session, we turned her loose in the pasture, and, to our amazement, she promptly started rolling on her own. We could actually hear the vertebrae popping into adjustment as she rolled for the first time. She made a rapid recovery, and she now drives although she is still unable to be ridden for long periods of time.

Not long after I began communicating with animals long distance, I started receiving letters from people asking, "If you can talk to animals at a distance, can you use ESP to find lost animals?"

Ten percent of the time I can. The rate of success is low because I see only through the animal's eyes. This "view" is often so general that it is extremely difficult to distinguish one terrain from another unless I have personally been there. Also, sometimes an animal may convey where it has been but not where it actually is now. I don't know why this happens, but it does.

One dramatic success at finding a lost animal occurred in 1974 while I was working in Cincinnati, Ohio. On December 1, Herman Koopman, a horse trainer who lives in Portola Valley, near San Francisco, telephoned me about his lost Arabian mare. (She had been trained for dressage, a fancy form of riding that takes a horse years to learn, like the Lippizan horses of Austria.) I had once visited Koopman's ranch, and so I recognized the area when I was finally able to "tune into" his mare and hear what had happened. She told me she had strayed from the rest of the herd while they were heading up the winding road to the pasture at the top of the hill. She saw that the other horses were above her and tried to take a shortcut—straight up the hillside. In the process she slipped and fell back down the hill, seriously injuring her left shoulder. Here she remained because her pain was too great to attempt climbing the hill. An hour after we had hung up, Mr. Koopman called back to tell me he had found the horse at the foot of the hill. Her shoulder was cut and bruised, and her left front leg was sprained. However, the mare eventually recovered, and Koopman was able to continue exhibiting her for dressage.

A call came from Julie and John Crawshaw of Pasadena, California. The Crawshaws owned a jet black cat called Jasper who suddenly disappeared. I "tuned in" to Jasper, and he showed me where he was—lying under the house next door. He had eaten something while outdoors that soon made him very ill, possibly poisoned. The cat felt he was going to die, and so he crawled under the house. I knew it was next door because I had been to the Crawshaw house before and recognized it through Jasper's eyes. The owners found Jasper lying under the house. Then I asked Jasper why he didn't go home. He said he knew he was dying. Many animals will leave home to die alone. I do not know why. The Crawshaws took Jasper to the veterinarian, but unfortunately he could not be saved.

One of the most gratifying cases I've had locating a lost animal was a Poodle named Gigi in San Filipe, Mexico. This is terrain I do not know. The Poodle's owner, Ann Fischer of Los Angeles, thought that Gigi had been stolen while she and her daughter were out boating. On their return the animal was gone—and so was the family camping next to them. I contacted Gigi mentally. Apparently she had started out for the desert, then decided to return to the campsite. On her way back she was captured by a Mexican family. Gigi's mental images showed her tied up outside an adobe hut with chickens near. Ann Fischer returned to the San Filipe area, posted a reward for Gigi's return, and the Mexican family responded the next day.

Another successful case was finding a Labrador which had disappeared with another dog. When I "tuned in" to them, they showed me a housing pen. They told me they had pushed open a gate and run across a well-cared for field with short grass and then into a farm. They were sniffing around the calves in a pasture when the farmer saw them and penned them up. The owner soon recognized the area. Next to the housing development where he lived with the dogs was an open field—a golf course. Across from the course was a farm. The owner called the farmer, who said he had confined the dogs because he was afraid they might harm his livestock. This took place near Asheville, North Carolina, another area unfamiliar to me.

I do not take lost animal cases any more because of the deluge of calls I started to receive—as many as twenty a day. I was spending five to six hours a day doing just this. Since I do not charge for this service, it prevented me from making a living. Now I work only through referrals from Pet Finders, Incorporated, and only on extremely rare cases. This organization, of which Jody King is president and I am vice-president, locates lost animals and whenever possible returns them to their owners. Miss King and I are also building a free diagnostic clinic for animals just north of Los Angeles in Ojai, California, and a retirement home for the pets of people who have died, a place where the surviving animals can peacefully live out their days.

Long distance ESP is more tiring than communicating with your animal in the same room, but it *is* possible, and as these few cases I've mentioned show, the results can be heartwarming.

6

IDENTIFYING EMOTIONAL PROBLEMS

Like humans, animals have emotions which may run the gamut from joy to grief. They also have emotional needs which, if deprived, may produce emotional problems. The way an animal expresses these problems, however, is very different from the way a person would. The healing process, on the other hand, is not so different. Understanding and communication are the two essentials.

Medicine today tells us that nearly 85 percent of all physical illnesses have emotional roots. This is an alarming percentage. Dr. Arnold A. Hutschnecker, author of *Will to Live,* is emphatic in his belief that the human mind can actually cause death. This psychotherapist contends that the "will" to live can end or cease, and when the subconscious knows and accepts this fact, the body reacts and physical death becomes a reality. Animals are also influenced by the "will" to live. From long experience and close observation, I know that the emotional quality of an animal's life and its physical well-being are directly related. The animal may receive excellent physical care, but if the outward physical symptoms are rooted in an emotional disturbance, healing will not occur. In some instances, death may result.

There is one difference between humans and animals that I should mention. In a human, the physical disorder which originates in (or is aggravated by) the psychic or emotional processes is considered psychosomatic. A human being, faced with a difficult life situation or crisis, may develop a physical illness—headache, nausea, and so on—which will reflect the tension. Animals do not have psychosomatically induced physical problems, but they *do use* physical symptoms as signs or signals to express emotional distress. And this is what to watch for.

Each pet's life is different. Only you, the owner, are capable of evaluating your pet's environment, noting day-to-day behavior changes, eating patterns, and any physical abnormalities which may or may not be signs of emotional problems. The first step, if you notice any unhealthy change, is to take your pet to your veterinarian for a complete physical examination. Do this before concluding that your pet has an emotional problem. After a clean bill of physical health, *then* look into possible emotional roots. Animals brought to me for consultation are those which exhibit unusual behavior changes or suffer from physical symptoms of distress or serious illness with no apparent reason and no verifiable medical cause.

Near-death and recovery was experienced by a race horse, Eagle's Dynasty, owned by Mr. and Mrs. Darrell Clingman of Arcadia, California. Eagle suffered a leg injury when he got cast in his stall. This ended his racing career. Several veterinarians were called in on the case, and their joint prognosis was dim. All agreed that Eagle would never run again, much less race. Some even doubted he would walk. Mrs. Clingman, accepting this no-recovery diagnosis, traded Eagle for a jumper. The new owner decided the horse could be used for breeding but kept Eagle at the Clingmans' farm for physical care during his convalescence. Eagle's condition worsened. After two months, the new owner decided it would be best to put the animal to sleep. At this point Mrs. Clingman, who still cared for Eagle, called me. She hoped that Eagle's life could be spared. I talked to Eagle for an hour, asking him what was wrong and what was happening to make his physical condition so much worse. Finally Eagle lay down, placed his head in my lap, and mentally transmitted his message. He wanted his former owners, the Clingmans, back. He felt the Clingmans were the only people who loved him, and he wanted to stay with them. He knew that once he got well he would have to leave them. He would rather die than go to a new home.

I conveyed the horse's feelings to Mrs. Clingman. As I did, Eagle raised his upper lip, quickly shook his head up and down, and whinnied in approval. The next day Mrs. Clingman retraded the jumper for Eagle. The horse's condition immediately started to improve. Eagle began to eat properly, and his leg injury began to heal. Since the Clingmans and I are now friends, I've seen Eagle's progress. He not only walks, but he runs in the pasture alongside the beautiful foals he has sired since his recovery. Eagle cannot race again, but the Clingmans still use him for breeding.

When a mysterious physical symptom surfaces, or healing seems impaired, never overlook the emotional quality of the animal's life. Eagle's Dynasty is a prime example because before we knew of his fear he had had the best possible veterinarian care.

Animals have also told me that they are affected by their owners' lives and *their* emotional states. An argument between two people may upset an animal because he feels the anger and cannot discern its direction. Pets will automatically blame themselves as the cause of the anger and react to this frustration by scratching or chewing at themselves. This is easily checked by observing a pet during a domestic quarrel. Frequently the animal will hide. Few humans understand what's going on during a heated discussion, and tempers are lost along with logic and reason. How is the animal to understand what we don't understand and are unable to handle logically? Animals tell me that during a domestic quarrel they need to know that they're not the object of anger and that it's not directed at them. Positive statements to this effect are helpful. There is no need to relate the whole argument. Just tell the animal that *it* is all right; then it can relax.

A consultation with a Dachshund, belonging to a friend of

mine, Wanda Selling of Covina, California, revealed just how damaging human conflict can be to a pet. When I first saw Liebchen, her body was covered with a skin rash, and she constantly chewed at her feet. The veterinarian on the case diagnosed a grass allergy. He advised confinement to the house and prescribed medication. After one week, the animal's skin condition worsened.

I asked her what was happening. She mentally communicated that the owner and her husband argued about problems in the home as well as the veterinarian's cost for skin treatments. The Dachshund automatically blamed herself for everything. Each time the couple argued, she chewed at her feet and scratched her body. Confinement heaped yet another frustration upon her. She missed running in the yard and the company of the dog next door. This longing and lack of exercise intensified her compulsion to scratch and chew. Based on what the pet told me, I advised the Sellings to increase her exercise and reassure her that family quarrels were not her problem. Within two weeks, Wanda notified me that the skin condition was greatly improved. It did not clear up entirely, and I am reasonably sure that this is because of the dog's sensitivity to domestic turmoil.

When we humans face frustration, we can go for a walk, eat, do something to divert ourselves from the problems and release tension. Animals do not have these options. They cannot escape their environment or give vent to frustration by tearing at whatever is around them because this behavior means punishment. The alternative, then, is to direct the frustration inward, to *tear at themselves;* scratching and chewing result. Domesticated animals need plenty of exercise, often a wonder cure for the ailing, urbanized pet. Owners sometimes believe two or three short daily walks on a leash is enough for a dog, but most dogs require far more. Apartment living is tolerable for pets, even large ones, provided the exercise requirement is fully met.

Otherwise excessive chewing and licking of the feet will probably result.

The most common physical symptoms of an emotional problem are hot spots and rashes. Rashes are skin irritations that cover the entire body. A hot spot is one point on the body where an animal chews until the area becomes raw and inflamed. In northern California the most common reason cited for skin irritation is fleas. In southern California it is blamed on the grass. From what the animals tell me, I believe the real reason is more often emotional, linked to the amount of time a pet spends in confinement or to some emotional frustration it cannot solve.

Princessa helped me understand this. She was accustomed to running free on the ranch in Duarte. Whenever she was in heat, I confined her to the house. Within a few days of the confinement, she started scratching herself until a hot spot appeared on her rump that remained until she had her freedom again. This was before I recognized and understood about animals' emotional needs and the symptoms of emotional problems.

Another example of the connection between confinement and skin rash is illustrated by a German Shepherd, belonging to Mr. and Mrs. Beckett of Sacramento, California. I was speaking at a club in Sacramento when the Becketts contacted me. Their dog had been plagued by skin rash most of his life and at one point had been taken to the Davis Veterinary Teaching Hospital in Sacramento. The doctors there located some staph infection but not enough to have caused an irritation as extensive as this one. The staph infection was cleared up, but the dog still suffered the skin problems. During consultation Thor conveyed to me a whole chain of emotional problems dating back to early puppyhood. Now confinement was adding a physical frustration to the emotional ones. After some work with Thor during which I helped him resolve his emotional conflicts, I advised his owners to lift the confinement and let him loose in his fenced yard. After this, whenever Thor started to scratch, a member of the family took him out running alongside a bicycle as a way to relieve the energy build-up. The owner called me two weeks later to report that the animal's skin problem had finally vanished, and upon further checking, a year later, it had not returned.

In identifying emotional problems, an owner must explore several avenues before he can rule out a purely physical cause. Increasing the amount of exercise is one way to eliminate skin disorders. Another is to evaluate the bathing and grooming schedule. A rash may be caused by bathing a pet too often. Veterinarians tell us that a dog should not be washed with soap. Soap strips the skin of the natural oils designed to protect the animal from dryness and from picking up bacteria or fungus infections. If a dog's diet is proper, all it needs is rinsing with clear water and frequent brushing to remove excessive dirt or an occasional bath with a shampoo obtained from your vet or pet shop that is either medicated or PH balanced for your dog. The human PH is different from that of animals. I wash my dogs with soap once a year, but they go swimming at least once a month. This plan insures healthy coats and a lack of "doggie" odor. If a dog has dry skin from overbathing or lack of oils in its diet, it will scratch. This will cause the skin to become irritated which, in turn, will make the animal vulnerable to infections. Fleas also flock to an irritated area, causing further irritation and greater itching. It becomes a vicious cycle for the poor dog.

Do not discard the possibility that improper diet may be responsible for what appears as a behavioral or emotional problem (see chap. 11).

One of the most common and unpleasant expressions of emotional problems is urinating or defecating in the house. Animals tell me that when they do this they are trying to eliminate something or somebody from their lives. Animals can experience personality clashes. I learned this from a Poodle who told me that every time the daughter in the family came home from college, he became upset and urinated on her car tires, suitcases, and shoes. The dog explained that he just could not tolerate the young woman and wanted her eliminated from the home. In another case, a Keeshound was brought to me for consultation because each morning he defecated on the wife's side of the bed. The dog mentally explained to me that he hated being in the dog shows, and the wife insisted on showing him. He never defecated on the husband's side of the bed because the man had often told his wife and the dog that he felt the dog should stay at home. Incredible? Not to that Keeshound!

Sometimes this urinary behavior crops up when two animals are feuding because of a personality clash. Two cats, Nupkins and Puffins, had a real battle going between them when their owner Marjorie Sutton called me. Mrs. Sutton said the cats were not getting along and one of them was urinating on the drapes, but she did not know which one. When I communicated with the cats, Nupkins readily admitted she was responsible. When I asked her why, she said the owner changed the litter brand from clay to one of the green ones. The cat did not like the smell of the litter, and it stuck to her paws. When Nupkins was finally forced by necessity to go to the litter pan, a grumpy Puffins who did not like Nupkins would be found standing in the way to make problems for Nupkins. Nupkins told me she found it was easier to use the drapes than confront Puffins in a fight for the pan. The owners purchased another pan and placed each pan in a separate room. They also returned to using the previous litter brand. The urinating stopped. Puffins also had a complaint which I conveyed to Mrs. Sutton. Apparently she had taken away the dry food Puffins usually snacked on during the day. This made him angry. Mrs. Sutton gave the cats separate dishes with plenty of dry food, and the two cats then tolerated each other.

Jealousy will also cause bad behavior. A Scottie, Rowdy, had this problem. His owners, Mr. and Mrs. Edward L. Coughlin of San Gabriel, called me because he had just been brought home and had immediately started urinating on Bootsy, their older dog, and on Bootsy's toys and basket and elswhere in the house. When I mentally communicated with Rowdy, he explained that he did not want to stay with these people. He had loved the kennels where he had his own run and where nobody showed partiality, and he wanted to go back. He was also jealous of the older dog, to whom he felt the family showed partiality. Rowdy's behavior was an expression of jealousy and hostility. The urinating stopped when the family made a point of treating both dogs equally. A deliberate treatment was initiated in order to help the animal make the new home adjustment. To cure Rowdy of his desire to return to the kennel, the staff was told to ignore him on his return visits for grooming. On previous trips, the staff fussed over him and Rowdy basked in the attention. But on the next visits, they groomed him with no fuss, placed him in a crate, and kept him there for the day. After two such visits, Rowdy was content to go back to his new home where they now lavished attention on him.

Behavior changes can also be brought about through mental projection coupled with the laying on of hands. This method is a reassuring way to transmit much-needed understanding. I have found the procedure extremely helpful, and it is an important and innovative aspect of ESP healing.

The idea was brought to my attention by one of England's top horse-trainers, Barbara Woodhouse, when I read her book *Talking to Animals.* Miss Woodhouse developed the method but does not tell what she was thinking. I feel reasonably sure she uses thought projection. This is mentally telling the animal she is its friend and would like it to be calm and gentle. I use this idea and also project feelings of peacefulness and gentleness.

An early experience I had with the laying on of hands, coupled with thought projection, came during a consultation with an unmanageable, nervous horse. Whenever he was touched on the rump or back legs, he balked and kicked. This made showing him almost impossible. Holding my two hands together, palms down, I began sliding them down along the horse's neck, gently pushing against the muscles. During this physical process, I simultaneously projected toward the horse a sense of quiet and relaxation. Pausing, I held the pressure firmly and still projected the thought patterns. I then slowly moved my hands along the horse's body, pausing for pressure, sliding slowly, then pausing again for pressure and gradually progressed to his hind quarters. The horse held still. Then, slowly, he lowered his head and half-closed his eyes as I continued to move my hands around his hind quarters and down his legs, encircling them and gently pressing inward. After four of these treatments, anyone could handle the horse. He had been reconditioned by thought and touch. When using this method, always slide the hands; never pat or lift the hands to change positions. This action disrupts and stimulates rather than calms.

One of the most dramatic examples of this treatment occurred when I was working in a veterinarian's office. A man and woman entered carrying an injured cat whose cries sounded like a baby screaming in pain. The distraught owners told me they thought the cat's front legs were broken. We had to wait for the doctor, and so I approached the animal, laid my hands on it, and mentally projected the positive sensation of pain leaving its body and being replaced by quietness and peace. Slowly, like Eagle's Dynasty, the cat laid his head down and closed its eyes. I explained to the owners that their own emotional state was greatly contributing to the cat's fears. I showed them how to lay hands on their pet, and they took over successfully where I left off. The cat no longer screamed; he just rested quietly under the umbrella of their thoughts.

Animals know when someone is receiving their thoughts, and they exhibit some interesting reactions. For example, during consultations animals rarely exemplify the bad behavior for which I was contacted. Instead, they appear relaxed, almost as if they were falling asleep. Animals that normally run from people or even bite people will usually lie quietly at my feet while I am communicating with them. An exception is the cat which hid behind a flower pot to stare at me from a safe vantage point while it tried to figure out why it could see its own thoughts (mental images) in my mind. As a general rule the animals just relax and give their full energies to projecting their feelings to me.

Another common reaction among animals is their emotional response to grief. I was working in Bradenton, Florida, in January 1975 when a woman called about her cats and the radical change in their behavior she had noticed shortly after the death of her husband. Before his death they had been extremely affectionate; now they were aloof to everyone, including her.

When I asked the cats why they had changed, they told me they were sad and missed their owner, and that they felt alone, as if nobody understood their grief. I explained this to the widow, who immediately gave them sympathy and love. The next day she called to tell me the cats had returned to their normal affectionate behavior. I attribute this change to the fact that the cats were finally able to express their feelings. It may sound trite, but it is nonetheless true: These cats needed someone to talk to. They needed to be relieved of their sorrow and anxiety. Once this had happened, there was no longer a need for abnormal behavior.

Two important symptoms of an emotional problem are shyness and overaggression. These are both learned behaviors and are not genetically linked (see chaps. 8 and 9).

Summing up, the symptoms of emotional disturbances are:

1. Chewing of the feet or excessive licking of the feet. The animals are trying to tell you they are not getting enough exercise. Cure: more exercise.
2. Wetting or messing. They are trying to tell you they want to eliminate someone or something from their lives. Try to connect the defecating or wetting with a person or a situation that is upsetting to them, and resolve the problem if possible. If not possible, discipline the animal.

3. Rashes and hotspots. The main cause is frustration. Find out what is causing the frustration and resolve it. Also check the animal's diet, frequency of bathing, and the products used in bathing it and laundering its bedding (see chap. 11).
4. Shyness and overagression (see chaps. 8 and 9).

PART II

YOU AND YOUR PET

7

SELECTING YOUR PET

Selecting a psychologically sound animal is not a matter of luck. If you look at a litter of puppies or kittens or walk along a line of cages of animals up for adoption, you can tell next to nothing about the pet and how it will fit into your home (if it's for your family) or will adjust to its duty (if it's a work dog). So it is wise to take some time and a little trouble before purchasing an animal. You may also enjoy yourself in the process.

The following criteria will help the prospective owner, as well as the animal, and therefore the relationship: (1) the needs of the owner, (2) the place of purchase, (3) an evaluation of the psychological and physical condition of the mother animal, and (4) the pet's early environment. There is a vast difference between a home environment and a kennel environment although both can be excellent places from which to select an animal.

Pets for the Family

Bringing a pet into your home means you are seeking a relationship that will, more than likely, span a good many years. It will be with a living, feeling creature that will endure emotional trauma at the change of ownership. Be prepared to make the purchase at a time when you are able to plan on a long-term relationship and give some time to the new member of the household.

First, decide on your needs. Then study the breeds best suited to those needs. If you want a family pet, one which will provide companionship for your children (particularly small children), *do not purchase a small dog.* Children are rough on small animals, and this may lead to snappish, aggressive behavior in animals that have no other way of defending themselves. A large dog can handle the roughhousing of children. Find one that has been born and raised in a family atmosphere. The pet's early conditioning is important. Children's feelings are changeable; one minute they are happy, the next, sad, the next, angry. Only an animal that is familiar with children can cope with such volatility. For the family environment, consider the calmer temperament commonly found in the larger breeds, such as the Great Dane, Standard Poodle, Samoyed, Collie, Cocker Spaniel, Labrador, and German Shepherd. The last two are more aggressive breeds but also possess a temperament that is very suitable for children.

When you live in a small area, larger breeds are fine as long as they are able to get enough exercise. But even if you have a large yard, a big dog will usually just lie around the back door and only run when someone comes out to play with it.

If you live in a city apartment where sufficient exercise may be difficult, consider getting a cat. Cats make excellent apartment dwellers, and they are good companions to children; they also take relatively little care, are naturally clean if provided with a litter pan, and do not require walking for exercise. I have spoken to many animals that have been raised indoors or in captivity, such as in a zoo, and they have told me they are perfectly content. They do not know any other way of life, and since their surroundings are clean and comfortable and their food is provided, they are happy. The only time I have found discontent in a caged animal is when it has not been cared for properly or has previously been free to roam in the open or the wild. After this, confinement indoors will produce emotional frustrations that result in either skin problems (rashes or hot spots) or destructive behavior such as wetting, digging, or tearing furniture or paper.

Birds, hamsters, and other small animals also make excellent apartment pets for children. For the mature apartment dweller who doesn't want to walk a dog two or three times a day, the small breeds of dogs get plenty of exercise racing from room to room and are easily trained to use papers or litter pans.

Be prepared to protect your animal—in many ways. If you have a large fenced area, consider your fencing. A small dog may get out and get lost or killed by a car. As I have mentioned, I am vice-president of Pet Finders, Incorporated, an organization in Los Angeles that helps people locate their lost animals. Every day purebred animals are stolen from cars and yards. The theft of small dogs from fenced yards is all too common. They are easy to resell and do not put up a fuss or bite savagely enough to defend themselves from the thief. Most stolen dogs are then resold under falsified papers as guard dogs or as animals used to train fighting dogs. Dog fighting, as a sport, is against the law in most of the United States, but it is still a common illegal practice, particularly in California. Any barking dog is a deterrent to a house thief, but a small dog cannot stop a dog thief from stealing it when left out in a poorly protected yard.

Under no circumstances should any animal be allowed to run free. No matter how well behaved your pet may be, there is always that one time when it will see something across the street and will dash in front of an oncoming car. You also risk the animal pounds picking up your pet, or people thinking it is a stray and taking it in. In coyote country, coyotes will lure loose female dogs into their pack by mentally calling them to join them. They tease them into following them. They kill and eat all small dogs, female or male. They also use their female coyotes, in heat, to lure male dogs out to a spot where the pack can kill the male for food.

Last, a warning about keeping a dog tied up. I have met many dogs that are tied up for various reasons, and they have told me that this drives them crazy. Dogs are better off confined to a small area than being tied up although they can be tied up for short periods such as a few hours. It is the constant tying that they cannot tolerate.

I have more to say on where and where not to purchase your pet, but here's a word to the wise about the pet you want for family companionship. Find one that has been raised with a family rather than being born and raised in a kennel. There is one exception: If the kennel-born pet has been handled a great deal by people until it is ready to go to its new home when it is about eight weeks old, then it will be all right. But animals that remain in a kennel situation begin to relate to a *place* rather than to people for their sense of security. In a home environment, they find it difficult living with people, or they may find it traumatic to change to a new place. A kennel-raised animal is content in a kennel. A homebred animal psychologically cannot tolerate being placed in a kennel, away from close contact with human beings, and will develop all kinds of physical ailments.

Pets for Protection

If you are looking for a guard animal, consider the more aggressive breeds such as Dobermans, German Shepherds, Labradors, Giant Schnauzers, Rottweilers, and Boxers. It should be raised in a kennel or outdoor situation so that it does not become too dependent on the companionship of people. This will prevent the animal from being lonely during the long hours of guarding the yard, garage, store, or whatever area needs protection. The guard dog must be taught not to eat anything in its area unless you, the owner, say it is all right. This will prevent it's eating poisoned food someone may throw to it. The dog must also be spayed or neutered to prevent it's being led off guard by a dog of the opposite sex. Spaying or neutering does not in any way deter the aggressive guard dog from its protective duty.

It is also best to have at least two dogs at the same time. I am a single woman, and I travel by myself in a camper all over the United States, lecturing and counseling with pets and groups. I take three Shepherds with me all the time because I feel that anybody or any gang trying to harm me will probably be able to subdue one dog or even two, but it would have an awful time trying to overcome three sets of powerful Shepherd teeth. Besides, the dogs are company for one another while I am busy. If you work and must leave your pet alone for hours at a time, consider getting two of the same species, or even a dog and a cat, for they can become wonderful companions. This will relieve your guilt at leaving them alone while you are socializing after work. Animals do get terribly lonely, especially guard dogs that have very little human companionship.

Where to Look for Your Pet

Now that you have decided on the approximate size and type of animal you want, based on your own requirements as well as the space and time you can give this pet, the next question is where to look for it.

First, visit the dog and cat shows, preferably the all-breed shows. You can get a comprehensive view of all available breeds and the best animals of those breeds. Newspapers list the times and places of these shows near the section where pets are advertised for sale. A "Specialty" listing refers to one breed or one class of breed only. A "Sporting Dog Specialty" refers to dogs that are used for hunting; "Toys" are the small breeds, and "Working Dogs" are those used for pulling, police work, guard, and herding. The American Kennel Club (AKC) is the only acceptable registration organization in this country, and its book contains lists, descriptions, and photographs of the different breeds and their standards and specifications.

When you attend shows, walk around the rings to the breed of pet you are considering, and ask the breeders questions. They are usually happy to share their knowledge, and you will get to know the points to look for. Talking to several breeders will give you differences of opinion, which will also be useful. The reputable breeder will be willing to show you his or her stock and the mother of the pet you wish to purchase. He or she can also show you records of wins and the places or people that have also purchased from him or her. This will enable you to follow up on the reliability and health of the animals.

The breeder will also be willing to give you a guarantee against hip displasia, a bone problem most commonly found in large dogs, which should surface by the time the animal is one year old. The cause of hip displasia is disputed, but the effect is undisputed: The hip ball and socket do not fit properly, and the animal either becomes crippled or develops painful arthritis. It is hard to detect in the pup. It is found only by X-ray in the grown dog and then only by a veterinarian who knows how to properly position the dog for the AKC requirements. Not all vets know the exact requirements. The dog's parents should be registered by the Orthopedic Foundation for Animals, which means that they have been X-rayed clear and have good hip socket and joint placement. If this problem surfaces, the breeder should either replace the pet or refund the money or a portion of it. You should get a written guarantee before you purchase the dog. Many people complain about having to pay for a dog when there are so many free ones available. This is true, but the free pets may not have had proper care and nutrition. The average person cannot afford to give proper care to a pet he or she intends giving away. Free pets often develop problems that eventually cost a great deal more in veterinary bills than the purchase price of a guaranteed pet. Although an animal may have champions in its background, it is the immediate parentage that is important.

Responsible breeders also take care of the stock they produce. One example is the Samoyed breeders in the San Francisco and Sacramento area. They have formed a club to rescue unclaimed Samoyeds from the pounds and place them in good homes if the original owners cannot be traced.

When I, as a breeder, place a dog, the new owner must agree to return the dog to me if for any reason the owner cannot keep the dog. One bitch I bought back is now my hope for a future champion. Another breeder was criticized for taking back a Samoyed bitch who was eight years old. This is generally past the showing age. Wilna Coulter of San Carlos, California, did not care since she feels responsible for her breedings. Her dog, Star, became a champion and is still unbeatable at ten years of age. She is nearly perfect in conformation. The conformation does not develop for a year or so, so when you sell or buy a baby, you may have unknowingly sold your best dog or bought a champion!

If it is to be a family pet, give the children an opportunity to learn the proper care of the animal and understand what is involved in that care. A reputable breeder will give your family that opportunity. Children must be taught the proper way to handle the animal and to respect the animal's need for love.

Where *Not* to Buy Your Pet

Never buy a pet from a swap meet or a flea market. These are favorite places for thieves to unload stolen animals. You have no way of checking out the papers on these animals or the conditions under which they have been raised.

Be extremely cautious when buying from an ad in the newspapers. Be sure you have the privilege of going to the home or kennel and checking the animal's breeding conditions. Be sure the animal really belongs to the person selling the pet because this is another common way of unloading stolen animals. One family had their Irish Setter stolen from their yard. Two months later they saw a "For Sale" ad in the local newspaper for an Irish Setter. When they went to the address, they discovered their dog with falsified papers. So beware. Any honest person will give you the opportunity to contact the owners of the sire and dam listed on the papers so that you can be sure the animal is what it is represented to be.

Last, I do not recommend purchasing a dog from a pet shop. In my work with breeders all over this country, I have not yet found one that would sell their stock through a pet shop. Responsible breeders want to keep track of their dogs to be sure they are in good homes. When I place my pups, I go to the home and see if it is suitable for the pups' mental and physical well-being. I make sure the dog will not be used for indiscriminate breeding or for creating another puppy factory—an inhumane place where people breed bitches every time (or nearly every time) they come in heat, raise the pups as cheaply as possible, and sell them to pet shops or wherever they can make a dollar. The kennels are usually too small for the animals; conditions are generally unclean; there is no room for exercise; the ventilation is bad; and no care is given the animals. The mother suffers greatly, crammed into a small cage with a litter constantly hanging onto her. Such a condition existed in El Monte, California, for several years before it was discovered and put out of business. On the evening news our local television station showed eighty Poodles with litters crammed into small cages in a dark garage. Pet shops are the favorite sales vehicle for these puppy factories.

Another good reason to stay away from pet shops is that they never have the mother, and you have no way of seeing her or the conditions under which the puppies (or kittens) have been raised. Unless the pet shop has been in business a long time, your guarantee, if you can get one, may not be worth the paper on which it is written. If the pet shop is reputable, you will be given the breeder's name and address so that you can check on all the information you receive.

A couple of years ago there was a pet shop called the Swiss Chalet in Pasadena, California. It specialized in Saint Bernards. It seemed reputable, had been in business for some time, and many people purchased dogs from it at about three hundred dollars a dog. One potential customer asked for a dog from a certain sire. The store owner said he would produce a puppy from that sire within a couple of months. Apparently he had contact with the major breeders of Saint Bernards and could get whatever the public desired. When the pup was produced and the papers signed, the shop owner was found guilty of falsifying papers. That particular famous sire had been dead for quite some time. Needless to say, this owner too was put out of business.

Many pet shops purchase from the pet factories I mentioned above and also from people who are stuck with puppies and kittens that are not good enough to sell and cannot be given away as pets.

Warnings

Do not be fooled by the claim that your pet has champions in its pedigree and is therefore more valuable. Almost every purebred animal on the market has pedigreed champions in its lineage, a fact that means absolutely nothing. It is the present breedings and recent records of show and health that count.

Beware of the claims of a new registry. There is no such thing. The only dog registry in America is the American Kennel Club. I had an experience in a small town in Mississippi in January

1976 on my way from Florida to California. I drive my camper on my travels so that I can take my animals. On this day we needed some pet supplies, and so I stopped at a pet shop in a mall and, as I usually do, began looking at the animals there. The sign on one cage read eighty-nine dollars. The dog looked like a Poodle, but not purebred. When I asked the shop owner if the dog was a purebred Poodle, he replied that it was a rare new breed called a Cockapoo. This is supposed to be a cross between a Cocker Spaniel and a Poodle, but in reality nearly any breed that is crossed with a Poodle is called a Cockapoo. I asked him how he could charge that amount of money for a mongrel. A Cockapoo, Peekapoo, Schnauzapoo, or whatever is being crossed with a Poodle is a mix breed and is not registerable and never will be registerable. This shop owner told me that the breed was recognized by a special registry, and he had papers to prove it. He also claimed that this dog was a real bargain. Breeders in other states were selling these dogs for one hundred fifty dollars. I knew this was an outright lie.

Run from a shop owner or breeder who crosses or mixes his stock, and check out all claims of this nature, either with the AKC or with legitimate breeders. Purchasing these animals at inflated prices only fosters continued breeding abuses. In my counseling, I find that the worst traits of the two breeds often combine to produce psychological problems. I personally have found few Cockapoos that are able to develop the deep sense of loyalty which makes dogs faithful pets. Many of them are literally untrainable because they are mentally retarded. Many are impossible to housebreak and are destructive.

Now that you have decided upon your need and examined all the possibilities to meet that need, you are practically ready to buy your pet. This is the final step. Get to know the mother of your desired purchase and the early conditioning of the offspring. An overwhelming majority of problems come directly from the mother and early conditioning. Many people have told me they have bred their male in order to get an animal with the same personality, but in reality all they got was fifty percent of the genetics, such as physical appearance and mental ability. The personality is formed by the mother and the environment. If you desire an animal like your male, get an approximately eight-week-old baby that was whelped by a mother just like your pet's mother, bring it home, and let your male finish raising it.

In this chapter I have emphasized breeds and, indirectly, purebred dogs, cats, and champions. I do not want to put down the mongrel. As a general rule I have found that, other than the Poodle mixes mentioned above, mongrels can make loving pets. If you can't find a dog or cat show and you don't care about the mix breed, then the best thing to look for is the healthy mother and the happy environment of her young. No matter what the breed, you want a psychologically sound pet.

8

THE PSYCHOLOGICALLY SOUND ANIMAL

For centuries humankind has believed that animals live strictly by instinct, but I believe, from my experience and from recent studies in imprinting, that little of what constitutes animal behavior is instinct. Animals learn from their mothers. This process starts at birth and continues in the case of dogs and cats beyond weaning and until about the time they obtain their independence and are removed from their mother's presence. Foals learn from their mothers until they too reach maturation.

The Mother

The pet's mother is all important: her health, both physical and mental, her environment, her prenatal attitude toward her litter, and last, the training she provides her young.

The mother's physical health depends on good nutrition. In the case of unborn dogs and cats, cartilage becomes bone in the last three weeks before delivery. Therefore the mother's diet should be supplemented with calcium, such as bone meal, with Vitamin A and D added so that the calcium can be absorbed. I also use a Vitamin E supplement to help strengthen the mother to carry her litter. This helps prevent spontaneous abortions. I use a Vitamin C supplement to help ward off any infections that may attack the mother before and after the birth. I use Brewer's Yeast to help build blood and energy, and I use alfalfa to be sure the mother has a good balance of minerals.

It is also extremely important that the mother animal have her innoculations against distemper, hepatitis, and leptospirosis since these diseases can be brought in by visitors. If anyone walks across an area where a diseased animal has been, the germs can be picked up on clothing and transmitted to the uninnoculated animal. The newborn puppies carry the immunity from the mother for about eight to twelve weeks if they are allowed to nurse her within the first twenty-four hours after birth, according to my veterinarian, Dr. Howard Kurtz. You should contact your own vet for the innoculation schedule.

Birth and Growth

From the time puppies are born, socialization with their littermates and with humans is essential for good mental and emotional growth. From the age of four weeks until they go to a new home, it is important to expose them to new situations. For example, I hope to show my pups, and this will require a great deal of traveling. So every day I take the pups for a ride to some new place. Since nothing upsets the mother dog, we visit the beach, the forest, parks; we watch fireworks, road construction, city traffic, and every situation I can find that they may have to face later. They get very excited at each new experience but quickly learn to relax and enjoy themselves as they watch their mother enjoying herself. When my pups go to their new homes, they are ready for whatever conditions they must live under. Obviously I do not allow my pups to go to a home where they will be placed in a back yard and left to themselves. You can purchase animals that are raised in kennels for that purpose.

When animals go to new homes, their emotions are still delicate; so they must continue to be exposed to new situations, or the early training will be lost. Pups and kittens should never go to a new home away from their mother and littermates until they are about eight weeks old. I have found that the mother is still training them and teaching them to think and react to life as an animal must, even though she has weaned them. When taken before that age, they become shy and insecure or overly aggressive. Puppies should be leash-broken by the age of four months; and they should start to know their basic obedience of *come, sit, heel, fetch.* But the training should only be for a few seconds at a time because the attention span is extremely short at this stage. One four-month-old dog was pushed by her master for ten minutes or so at a time. She became so nervous she stopped eating and started throwing up; she scratched excessively and tried to hide when she saw him coming. When dogs are six months old, they can master their basic obedience, but training time must still be limited. Flightier breeds such as Samoyeds, Shelties, and Irish Setters have such a short attention span that they can handle only a very few minutes. More settled breeds will learn faster, but you should still limit the sessions. Serious obedience or attack training or retrieving training should not be attempted until the dog is at least a year old or nearly two years old. The animal does not fully mature emotionally or mentally until it is nearly three. This is why a female dog should not be bred until she is psychologically mature enough to settle down and train her puppies to be well-adjusted dogs. The mother dog needs time to develop physically too.

Many people complain that the mother animal will not allow anyone to come near her young but tries to "protect" them. I do not believe this. If the mother animal loves and trusts you, the owner, what is there to protect her babies from? I have found from talking with mother animals that they are really jealous. Until the birth of their litter, they have received a lot of attention, especially during pregnancy. Suddenly everyone is making a fuss over the little ones and ignoring the mother. When Princessa had her litter, everyone who came to see the puppies was instructed to first make a big fuss over her. They were to praise her and tell her what a wonderful thing she had done. After about two or three visitors, Princessa met people at the door and even led them to her puppies so that she could get a great deal of praise and petting. She never once showed any concern over the pups, joyfully allowing people to handle and cuddle them. The only time she showed any concern was when a stranger started to carry one of the pups out of the room. Princessa trotted along after the stranger, without showing any aggression, just to see what was happening to her puppy. I made sure that nobody harmed her puppies, and she knew it; so there was no need for protection or jealousy.

It has been a beautiful experience watching her daughter, Philea, with her first litter, which was born not long ago. She does not trust strangers as Princessa does; so I made sure that plenty of people, particularly children, were present during the birth and immediately afterward. Philea tried to show aggression, but it was curbed, and she now accepts visitors and their praise. After this, I got some children to help stress-train the pups (a procedure I describe later), which proved to be a growing experience for both animals and children. Princessa was present during the births of Philea's pups and even helped clean them up. Philea's brother, Loverboy, has enjoyed sniffing them and trying to figure out what is going on. The cats have also investigated them, as well as my sixteen-year-old male Pomeranian, Blacky. It is a beautiful sight to see the mother, Philea, sharing her family with the rest of our animal household.

Stress-Training

Stress-training is most important—from birth to maturity. It is a way of conditioning the newborn animal's nervous system so that when it gets older and must face stress situations it will be able to withstand pressure. The Air Force stress-trains its military dogs because the stress-trained pup develops into a dog that can perform better for longer periods of time. According to an article in the *Los Angeles Times* May 3, 1970, "Top Dog in the Army,"

> Superdogs are also taught alertness and obedience, and ability to relate to their handlers. Among the tests are a pup's reaction to a rag waved at him; ability to escape from a maze; reaction to noises, lights and mirrors; alertness to hidden decoys; and ability to read hand signals. Training begins at birth and continues until the age of eleven months when the animals that pass these rigorous tests are shipped out for active duty. Four-day-old dogs are put in a centrifuge for a three-minute stress test.

What does this have to do with you and your pet? Everything! A new home is a stress situation. So also are traffic, strange visitors, fireworks, horns honking, another pet of a different species, being put in a kennel while you are away, children, not to mention a childrens' birthday party. You may wonder what a centrifuge is and whether it is necessary. It isn't. Handling does practically the same thing. Here's what I do with my newborn pups. On the first day of the pup's life I pick it up and stroke it. The second day I repeat the stroking and then slowly turn it around in a circle. The third day I repeat the first two days' procedure and then hold the pup upright on end, head up and rear down. The fourth day I reverse this position and hold it head down and rear up. But only for a few seconds at a time. Each day I continue to add a new position to the routine. I also increase the amount of handling time. By the time they are three weeks old, each puppy is handled for a minimum of ten to fifteen minutes per day.

Learning Independence

From birth to a certain age, depending on the species, the mother animal is teaching her young how to survive. This is why it is important not to adopt a too-young pet or give a too-young pet away. The mother dog or cat weans her litter when they are between four to five weeks old. The foal is weaned by the time it is four or five months old but should not be taken away from the mother until it is at least six months old. Puppies and kittens should stay with the mother until they are about eight weeks old because the mother must teach them how to think and how to react to the world as her species must react to survive. When an animal is taken away from its mother too soon, it gets confused because it is learning from a substitute mother. (I shall refer to this later in this chapter: imprinting from a substitute mother.) Then, as it grows older, the animal has difficulty relating to its own kind. In many instances it will refuse to mate. If it does mate, it may react in a fighting, aggressive manner. An animal taken away before proper development time may also become shy or insecure and develop any number of neurotic symptoms relating to trauma and early frustration.

The advantages of leaving puppies with their mother for eight weeks is beautifully illustrated by Princessa. This Shepherd weaned her pups at five weeks, but every day she would take off to the woods with them trailing behind, their tiny, round bodies bouncing up and down, feet in a furious scramble to keep up with her. Usually Princessa kept them out for only half an hour, and the pups returned in the same gleeful spirits they had left. But one day, she and the litter disappeared for more than two hours. I frantically searched for Princessa and the puppies, terrified that something dreadful had happened to them. Later, I sighted Princessa trotting down the road, her head held high, a look of triumph on her face. Dragging behind her, strung out in a row were seven soaking wet little puppies, so tired they could hardly walk. When I asked Princessa why she had taken them away, she told me that she was teaching them how to hunt. I have learned so much from Princessa and feel such gratitude for her forthright communication with me. There is absolutely no substitute for the education a mother provides her young during the first eight weeks. I have exposed Princessa's puppies to every new experience I could find, including fireworks, and still the best lessons are those they learned from their mother. They often become excited or afraid of a new situation until they notice her reactions. Seeing her so calm, taking everything in stride, even enjoying noisy fireworks, they copy her behavior.

One important lesson was with her son, Loverboy. For his first year and a half, he had little to do with children. This lack of exposure makes most dogs timid and untrustworthy around children, but Princessa made up for Loverboy's disadvantage. Princessa had been raised by children and played with them until I purchased her. When children were around, she played with them, fetching sticks and bounding about with them. Lover watched, perplexed that those jumpy unpredictable little creatures could be such fun for his mother when they were so changeable. I stood aside, reading his mind. Soon I saw Lover could stand it no longer. He approached the children. Then slowly, when he discovered that these children loved to play as much as he did, Lover became what I can only call "kid crazy." Now if he sees a child outside, he begs to be let out. When Lover runs with a child, it is as if his spirit has been freed. I feel his joy and his capacity for loving a child, especially a disturbed or handicapped child. Without question, Princessa had done a good job.

Through my work and Princessa's teaching, I have discovered that puppies that are still living at home should be taken away from the mother from time to time—from age eight weeks to one year. This will prevent the mother from dominating the pup and controlling its actions and thoughts, which, in turn, will result in an insecure, submissive animal. Puppies must be exposed to situations without the mother so that they learn to think and act on their own. This is not as important with a strong-minded male because he has a natural aggressiveness which will enable him to assert himself. But watch for any signs of shyness or submission in the pup toward its mother. Investigate the mother of the animal you intend to buy, and guard against buying a pet with an overly protective mother. She can be easily identified by her effort to keep people from getting near her offspring.

A thoroughbred yearling colt in San Diego suffered tremendously from his experience with his mother. I was called by the owner to find out why the yearling backed people into a corner of his stall to kick them and why he refused to allow anyone to put a halter on him. He also turned his head away when somebody passed, and he frequently walked with his head down, making strange motions with his mouth. The owners could not understand what was happening because they had never abused the animal.

I asked the colt what was wrong. He communicated to me, visualizing it graphically, that his mother hated racing, and she had told him people would take him away from her and force him to race. He said his mother told him to reject all people. That was why he turned his head away or tried to back people into a corner of the stall and kick them. I told this to the owner. She then recalled that every time she had gone into the pasture the mother had stepped between her and the colt.

This colt also told me that when he was very young he and his mother were transported to another farm. When the new owners took the mother out of the stall, the colt panicked so badly they put a halter on him and tied him to the stall. When his mother was taken away, he reared violently and tangled his legs in the lead rope; this, in turn, pulled his head down, forcing his nose against the wall. He ended up on his haunches with his legs in the rope and his nose pushed against the wall for quite some time. As he fought to get up, he rubbed raw spots on his rump. I visualized this experience and placed the colt's age at about three months. The owner told me that this was the time the colt had been transported to the farm where the mother had been bred. The owner said she oftened wondered where the colt had gotten the sores, but the breeder did not mention the incident. I firmly believe that if this colt's mother had not been so overly protective none of this would have happened. The trauma in this yearling was so great he was still functioning as a three-month-old foal. He had stopped growing emotionally at the time of the trauma. The motion the animal made with his mouth and the lowered head was actually a reenactment of the nursing position. I suggested a nursing bottle to ease the desire to nurse. I also taught the owners how to visualize to him the actions they wanted and expected from him, such as allowing them to enter the stall and handle him with safety. They were also instructed to promise to take him out into the pasture again (where he very much wanted to be) as soon as he allowed them to place a halter on him and lead him about. Five days later the owner called to tell me that the colt had responded to the therapy. He was allowing them to handle him and lead him. After the nursing bottle was offered, he quickly rejected it and stopped the head-down, nursing motion. He was then well on his way to racing. I felt this was good, for he had the competitive spirit which is essential in a horse that will spend his early life racing.

Observing and learning from these animals, I began to see a certain phenomenon occurring with regularity: the mother animal will teach, not only her good qualities, but her bad qualities as well. One such example occurred in Las Vegas, Nevada, where I was working with a dog club. Four different people brought dogs to me for consultation but did not tell me that the dogs were related. It seemed strange that four dogs would have exactly the same emotional problem and tell me similar stories about their mother. Finally I realized they were talking about the same mother. When I asked my host if the dogs were related, he said they were. They were not only related; they were littermates. I then met the mother dog and realized that she was very shy and insecure. Since the puppies communicated with her in their nonverbal intimacy, they had learned to think as she did, and also feel as she did: shy and insecure.

Imprinting

My personal experience with my animal family and also as an animal counselor are clearly responsible for my initial statement about animal behavior: that it is *learned* behavior rather than instinct and that it is learned from the mother because the mother is the first object the newborn animal feels, senses, and observes. This theory is reinforced by recent studies in imprinting, a relatively new discovery in the field of applied psychology. *Imprinting,* according to the dictionary, is to "impress" or "mark on" an idea "in the mind." Technically, in psychological terms, it is the discovery that the animal identifies the first object it sees as its mother and will thereafter copy the behavior patterns of that object, whether it is human, a like animal, or another animal species. Konrad Lorenz is the pioneer of this concept. I shall cite three examples. Two are about ducks, and one is about a dog and a kitten.

The first duck case was described in a film in my psychology and again in my biology classes and was a controlled experiment rather than an accident. A couple decided to experiment with ducks hatched in an incubator. The people waited for the eggs to hatch and were on hand as the first objects the new ducklings saw. The ducklings then perceived these humans as their parents and followed them about as they would a regular duck parent. They copied, as much as possible, the behavior of the humans. The humans even had to teach the ducklings to swim.

A similar case concerns a duck in Cincinnati, Ohio, where I appeared on the Nick Kluney television show, except that this duck's fate was an accident. When I appear on TV, the producers usually arrange to have local animals there for demonstration purposes—to show how I communicate with animals. If there is a problem with the animal, I ask it what caused the problem and then relay the information to the audience. One lady brought a duck to the show to find out why the duck laid her eggs in the winter, in the snow, and refused to mate or sit on the eggs. She, the duck, communicated to me that she felt the humans were her family since they had raised her in the house from the moment she was hatched. It is biologically correct for a duck to lay her eggs in warmer weather. This duck laid her eggs in the winter because she was in the warm house in the winter; she associated this with the proper warmth in which to lay her egg. However, she was often let outside to enjoy the air, and since she had to lay the egg somewhere and really didn't know what to do with it, she laid it outside in the snow or wherever she happened to be. She had been raised by humans and not by ducks; so when she felt the natural instinct to mate, she wanted to mate with a human and not a duck. She was utterly unable to relate to her own kind. Since she had been raised in a warm house, her biological timetable was also confused and off-balance.

Another case of imprinting which I find particularly interesting involves two animals of different species. About two years ago I was adressing a group of animal owners and counseling their pets in Omaha, Nebraska, when I met a lovely old French Briard dog. She was so old that her owners kept an old T-shirt on her to keep her warm. (Circulation decreases in the older animal as it does in the human.) This dog had rescued an abandoned kitten from the barnyard before the kitten's eyes were opened and then had proceeded to raise the kitten as if it were her own—by cleaning it and protecting it from danger. When the kitten opened its eyes, the first thing it saw was the dog. The kitten told me that she thought the dog was her mother, and wherever the dog went, the kitten also went, riding on the dog's back, clinging to the T-shirt. The kitten walked and acted like a dog, not a cat, because it had copied the dog's actions and thought patterns.

Now back to you and your pet. You have selected it and brought it home. You know its background and are pretty well prepared to meet its demands. Horse, dog, cat, as well as very young puppy and kitten, will need your attention and some very special love in the beginning. But does that mean spoiling it? You also have your needs. Your new pet is now a member of the family (unless you have purchased a guard dog), and there are some rules for it. As a pet owner and breeder I am sometimes considered a slave to my animals. This is not true. My animals both respect and obey me, and unless you want anarchy in your home, you'll want your animal or animals to respect and obey you too.

9

THE EMOTIONAL NECESSITY OF RESPECT

By following certain guidelines, you can have a long, happy life with your new pet, but not if your pet is lost. It takes so little to lose an animal: an unlocked gate, injury, wanderlust, chasing another animal, sex, and theft. For the pet's sake, take precautions. Every animal should wear an identification tag at all times, and it should also be tattooed. A tattoo makes the animal identifiable so that when a thief discovers the animal cannot be sold under false papers he or she will turn it loose. You will then have a chance of finding it.

The price of not tattooing your pet or not purchasing an identification tag can be painfully high. I once knew of a San Francisco lady who never allowed her dog to run loose in the streets or even walk with her in traffic. Since the dog, a Poodle, always stayed in the house or the yard, its owner considered a collar unnecessary. She also felt it would make the dog uncomfortable. One day the house caught fire while she was away, and when the firemen broke in, the panicked dog escaped in the confusion and ran several blocks. Since it had never been in traffic before, it did not know how to behave, ran in front of a car, and was hit. Nobody in the neighborhood recognized the animal, but someone did take the injured dog to a veterinarian for treatment. Without identification, however, the vet could not legally treat the animal unless he had permission from the owner or someone responsible. The animal died before the owner could be located. Such tragedies occur too frequently.

Care is also needed in the choice of a collar. Never leave a choke collar on an animal after walking it. These collars are dangerous. A pet can strangle if the collar catches on a sufficiently strong object. A dog may jump up near a fence, catch the collar loop on a wire or a nail, and die by hanging because it cannot slip out of the noose. For identification purposes I recommend a leather collar, loose enough for the animal to free itself without injury should it get caught. For cats I recommend a piece of elastic in the collar or a collar loose enough so that the cat can wriggle free if necessary.

Frustrations and Discipline

Pets can be fun, and life with them can be enjoyable. But they can also do a great deal of damage at times and "act up" in distressing ways. Most damaging behavior (whether to our home furnishings, other pets, or ourselves) and most "acting up" are symptoms of frustration or lack of discipline. In this chapter I will deal with some of the daily-life problems most pet owners face, as well as how to establish a good owner-pet relationship. Such a relationship is based, first and foremost, on the emotional necessity of respect, taught by enforcing discipline that is both reasonable and understandable to your pet. You may face problems of rivalry within the pet family, as well as the larger and more complex conflict between human and animal needs. You may also face problems stemming from animal insecurity, which may be expressed in either overaggression and attack or shyness and fear. Last, animals are subject to frustrations. The symptoms run a wide gamut and include barking, car chasing, furniture scratching and chewing, biting, hole digging, wetting, spraying, and other "antisocial" behavior.

One interesting fact that emerged from my growing awareness of my pets' psychological problems is the similarity of their difficulties to the behavior problems I encountered in my study of child psychology. Once I made this discovery, I was quick to begin applying to animals certain principles which had been successful with children, first with my own animals, and later with those I was called on for consultation. It is a deeply rewarding experience to help produce a harmonious relationship between owner and pet, pets among themselves, pet families, and their interrelationships.

Discipline leads to respect, and respect is essential to any relationship. This first lesson was taught me by several mother dogs as I watched them train their puppies. When respect is missing, the inevitable result is frustration, which surfaces in unacceptable behavior.

One day Honey Bear, a neighbor's long-haired black Chihuahua, was staying at my home with her three puppies. When she started eating, her puppies came racing over to see what looked and smelled so interesting, and Honey Bear snapped and snarled at them. I scolded her for not letting the puppies share her food, and she sent me a strong message to stay out of it. She was teaching her family to respect her. When I questioned the puppies, they admitted they were not hungry, just curious.

Some time later, when Princessa had her puppies, I observed her teachings, which seemed remarkably thought-out and premeditated. In one instance, the schedule took several weeks and considerable plotting. For about two years I watched Princessa bury her bones in the avocado grove, then apparently forget all about them. When the pups were old enough to be out in the pen, unattended by their mother, she proceeded to retrieve the bones and pile them in a heap, just outside the pen. She made no attempt to chew on them but just left them there in a pile. Later, when I let the puppies out for their romp, she lay down by the pile and started to chew. The excited puppies bounced over to her treasure and attempted to grab some of the bones. Princessa promptly attacked them. Terrified she would hurt them, I raced to protect the puppies, but Princessa emphatically told me to stay away. She had brought the bones there to teach them to respect her right to her possessions. Watching her, I realized she had planned the whole affair—to teach them a valuable lesson. Several times after that, when I was petting her and the puppies tried to interfere, she also showed a snarling reprimand to their invasion of her right to be loved. I allowed her to be dominant until the puppies reached four months of age; then I applied my own principles of equality to prevent the pecking-order problem.

In these cases the mother animal uses food and attention to teach respect. I began to realize that these were the two most important areas in the domesticated animal's life; when either of these is violated by another animal, fighting begins.

Three Rules for the Home

I have found that when I apply three rules to my animals' relationships it prevents conflict in the home.

First, each animal should have its own food dish which should never be violated by another animal under any circumstances. If there are leftovers you wish to give the other animal, put it in the pet's own dish, making the transfer when you cannot be observed by the animals. Even when one of my pets, particularly Loverboy who has a tendency to leave his food, sees another pet checking his dish, he doesn't show anger but asks me to make the other animal leave it alone. It is *his* dish. He has conveyed a hurt feeling when I have allowed this violation to go undisciplined.

I find it usually does not bother cats to share the dry food dish, but not at the same time, whereas the wet food or meat meal is precious to them. These should be fed separately and in different dishes. Horses also show the same possession about their grain buckets and hay racks. Water is not important, and none of them has ever complained about sharing a water bowl.

Each dish should be as distinct as possible in shape, size, and color. We are told by biologists that animals do not see color, but they communicate four basic colors to me: bright yellow, blue, green, and red. Some animals can even distinguish between brown and black, but orange, red, and pink all look red to them. And all pastel colors look white to them. I am referring mostly to dogs and cats.

The second rule concerns petting. When you are petting one animal, never allow another animal to push his nose in between the two of you. Many people think this act of jealousy is cute, but it is not; it is an intrusion and shows disrespect. Animals learn that by pushing each other aside, they get their own way. They convey to me that they are not aware of time, such as the difference between ten minutes and thirty minutes; so they do not care if you spend an hour with one and then maybe a minute or so with the other as long as each gets some attention. Time in this sense is a human concept.

When you are petting one animal, push the second one back and say (remembering to use positive statements), *You wait your turn.* When you set aside the first pet, give the second a little individual attention and then the third if that is the case. Princessa snarled at the puppies when they interfered with my carressing her, but when they were four months old, I began to teach her equality. I pushed the pups back and scolded her for snarling, by visualization as well as saying out loud, *"I am now the boss here. And I will make them respect you from now on."* She finally accepted her equal position, and we have no more problems. Sometimes when a pet sees two people embracing each other, it can sense the good feelings and want to share them, but it should not be allowed to push between and interrupt this affection because it could create jealousy and foster disrespect. It is best to curb this behavior because you cannot be sure of the motive. Very likely it is jealousy.

The third rule is that there is one boss in the home, one dominant figure, and that is the pet owner, not one of the pets. Animal dominance should not be allowed in any domestic situation. In the wild, where the survival of the fittest is important in order to preserve the best of the species, a dominant figure is necessary. This is not so in a domestic situation. I am often told that animals have lived together for years with one animal as the boss and there have been no animal fights. This may be true, but somewhere along the line I find that the dominated pet gets tired of being pushed around and will suddenly fight back. After having been bullied for so long its anger has built up so that the ensuing battle is ferocious and almost impossible to stop. When I am called in to counsel on this sort of problem, the animals tell me they are tired of being pushed around. Thus I usually find a dominating problem behind the quarrel.

People can fight with words, courts of law, and so on, but animals must fight with teeth and claws. It is just not worth taking a chance and allowing this fighting and rivalry for position to start. So stop it before it starts. You, the owner, are boss. Your pets are equals in your friendship and protection and love. Giving each pet its share is both fair and rewarding.

The second reason for insisting on animal equality in the home is to prevent the destruction of the dominated animal's personality. In almost every case, when the dominant animal dies, the dominated pet becomes totally different with a beautiful new personality. The dominated animal may also express joy at the death of the domineering animal which has made its life miserable for many years. You should never allow one animal to rule over another. The hurt is intense.

Hurt can be physical or emotional. The emotional hurt results in frustration, and the results of frustration are types of behavior which need various kinds of release.

Two of the most common frustration symptions in animals are chewing and digging. These are ways of releasing tension, and they need curbing. But before you administer discipline, you should first uncover the source of the anger or frustration and if possible correct it. It may be lack of exercise. It may be the domination of one animal over another, against which it cannot fight. Or it may be anger at you for not taking it with you on an outing or a trip.

One of the biggest mistakes many people make is based on the assumption that they must catch the animal in the act in order to administer discipline. This is not true. An animal can remember incidents in its life that date back to birth; it can also remember what it did a few minutes or a few hours ago. Take the pet back to the damage, visualize what it did, and tell it that is why you are disciplining it. Then go ahead and punish it. If you must leave your pets for long hours, be sure they have a clean litter pan (dogs use them too) or papers on which to relieve themselves. If your animal (either male or female) then wets, it is because of anger and must be punished. Unaltered males will wet for territorial rights, and excitement causes pressure on the prostrate gland which makes them feel sexually stimulated. They cannot help it. Only neutering will correct this wetting.

In any case, some favorite tantrums include overturning waste baskets, tearing or chewing papers, and pulling things off the bed or couch. If the cause of frustration cannot be eliminated, then the tantrum behavior must be controlled or curbed. Many dogs dig in the yard, looking for cool earth to lie in; so be sure they have one moist spot in the yard that belongs to them, and teach them to leave the rest of the area alone. Provide plenty of chewing or chewable toys to give the animals the chance for mauling and aggressive behavior; then when damage is done to the home, discipline the animals.

Cats pose three common problems: scratching furniture, spraying (males and females both do it), and eating house plants. I have asked the cats that live indoors all the time how they feel about being declawed, and they have told me that it does not bother them. They do not miss their claws, and when the veterinarian does a professional job, it is a relatively painless procedure. This declawing will solve the furniture scratching. Never declaw an outdoor cat because you take away its primary defense.

When cats spray your furniture, try to determine the cause, and then eliminate the source of anger if that is the cause. Wash the sprayed area with rubbing alcohol or vinegar to kill the scent. (Reasons for not using other disinfectants are given in chap. 11.) Place moth balls around the area where the animals wet or scratch; the odor will usually deter them from using that particular place again. The moth ball odor only lingers in the air for about twenty-four hours, but cats continue to smell it when they approach that place.

When cats eat plants, they are looking for minerals which are missing from their diets, as well as roughage. I suggest buying alfalfa tablets (not alfalfa sprouts) from any health food store, grinding a few tablets up, and mixing this in their food. The alfalfa plant leaf is the richest source of minerals I know and contains a complete balance of minerals from nature. If possible, grow a box of grass for your cats to eat. This roughage will also help.

I have followed the above rules for several years now, and with four dogs and three cats, I can honestly say they work. These are discipline, love, and dietary rules under conditions of training, breeding, showing, traveling, and of course our own kind of fun. People come into my home or see us traveling and stand amazed when they see all seven animals moving about freely, not divided by cages or doors or runs, the cats walking with the dogs or rubbing the dogs with affection, and vice versa.

Insecurity and Aggression

One of the most common problems I deal with is insecurity. This emotional root may surface as either overaggression or shyness. Insecurity usually has one of two causes. Either the animal has been deprived of exposure to new situations when it was young, or its mother, littermate, or some other animal friend has been allowed to dominate it. I have already discussed ways of helping the young animal grow normally. If, however, your animal has already developed this problem, there are definite ways to cure it.

Aggression is normal and good, but when aggression results in unprovoked attacks on humans and other animals, it is no longer healthy and is a behavior which must be unlearned. In my opinion, overaggression is a learned behavior. I have never found any evidence to prove it is an inherited trait, or "in the blood." The overaggressive animal must be taught, therefore, to curb its aggression and behave acceptably. By using a case history, I will explain, psychologically, how animals learn bad behavior.

A client brought a Labrador to me to find out why this normally mild-mannered breed was biting everyone, including his owner. The dog told me that when he was a puppy he didn't particularly like people. When people came to the home to visit, he showed aggression by snarling in a threatening manner. The owners were afraid he would hurt someone; so instead of correcting the puppy's behavior, they put him outside. The dog then realized that if he didn't like someone he could threaten or bite that person and get his own way, which was to be put outside and not have to tolerate people.

After four years this pattern of behavior became so deeply ingrained that the animal was helpless, and so were his owners. When he showed the overaggression, a groove was established in the brain patterns. Each time this was repeated and he got his way, the groove became deeper and more permanently set. In this particular case it would have taken a tremendous amount of work and time to undo what the family had, although quite unintentionally, actually *taught* the dog to do, namely, attack people. By the time I was called for consultation, the dog was already too dangerous to give to anyone, and since the family did not have the confidence or the ability to correct the problem, I recommended the dog be put to sleep. This was one of the few cases in which I have recommended such drastic action, for I find that most problems can be solved or corrected with effort, intelligence, and patience.

The first step in curbing overaggression is to get the dog neutered. In my experience the problem is more prevalent in males than in females. Neutering the male takes the edge off his aggression. Spaying the female, however, does not seem to help this problem although spaying helps other problems, which I discuss in the next chapter.

The second step is to make the animal behave. This can be done with an instrument of discipline, that is, something that is *not* used for any other reason around the animals. In this way they learn to identify the object with discipline. A warning: Never use your hand because then when you reach for the animal it will not know whether you are going to hit it or pet it, and it will become "hand shy." Do not use a newspaper either because then when you pick up a paper or magazine to read it the animal will run from you or become "paper shy." With my dogs, since I never wear belts, I use a wide plastic belt that makes a lot of noise but does not do any bodily damage. For cats, I find a squirt gun works wonders. In most circumstances I have only had to apply this treatment once or twice. After that, all I have to do is say, "I'll get the belt," and they know enough to behave. For small animals I have found that a soft plastic fly swatter is very effective and safe. All animals will run from a fly swatter because they know you are thinking of killing someone and they don't want to get hurt. Remember how the animal is able to read your mind?

Another form of discipline is tying an animal up when it has been bad. Most animals hate restraint. So let your pet know you are angry at it and that it is being tied up or shut out of the house until *it* is ready to ask forgiveness. Never approach your pet to forgive it, or it will feel it has gotten away with the bad behavior. Your pet must make the first overture. When asking forgiveness, animals usually let their tails droop, their heads go down, and they sort of sidle over to you. Then you should tell them you forgive them, but tell them they must be good or you will punish them again. Visualize the bad act and the act of disciplining them so that they know exactly what you are talking about.

If whacking them on the rump, firmly saying no, and using the other forms of discipline I have suggested does not correct the overaggressive behavior, then you will have to use the final, drastic measure—the shock method. This should *only* be used on the overly aggressive animal, never on a shy animal, and only as a last resort. I cannot emphasize this enough.

An animal becomes overly aggressive either because it has learned to be that way or because it is insecure. In both cases treatment or correction is the same. When an insecure animal attacks another animal or child, it does so out of fear and is called a fear-biter. (I am not talking about an animal defending itself when cornered by someone intent on hurting it but about the insecure animal that shows aggressive biting when it is cornered, perhaps by a child who only wants to pet it.) Insecure pets need to get on top of a situation before it gets on top of them. You may have to use a shock method to break the pattern.

The shock method is done by an electric cattle prod stick or shock collar (Cost: $6.50 to $200.00). The stick is used by cattlemen to make the animals move through shutes. The shock collar is used by hunting-dog trainers.

The cattle prod or hot-shock stick is a very handy gadget in the hands of a responsible pet owner. It looks like a long flashlight with two prongs on the end. It is activated by pushing a button on the side which causes an electrical current to shoot out of the prongs when contact is made with a person or animal. It is extremely helpful for people who walk their dogs and are bothered by other animals trying to attack their pets. When the button is pushed, it makes a buzzing sound. This is usually sufficient to stop any attacking animal. It is also very useful in breaking up a dog fight already in progress. Never attempt to push dogs apart with your hands because they will very likely bite you without even knowing it. Get one dog at the base of the brain with a good hold on the skin so that it cannot turn far enough around to bite you. Then push the buzzer on the prod stick and thrust the prongs into the neck or face area of the other dog. The electrical current will help "shock" him back to rational thinking. This will enable you to regain control and calm both animals.

The third use of the prod stick is best illustrated by an experience I had with Princessa. When I lived on a ranch in Duarte, California, we had several chickens. We leased the pastures to the local dairies; they kept pregnant cows there until they were ready to deliver. Princessa had been trained to respect the livestock, but the neighbor dogs, a German Shepherd and a Samoyed, had not been trained at all. They would slip under our fence and kill the chickens or chase the cows, causing them to abort their calves. We tried numerous ways to keep the dogs off, but without success. Finally we taught Princessa to chase every dog or coyote that came near the stock. This she did with great gusto. We soon learned though that she believed she was to chase every dog, whether on or off the ranch. I tried to teach her the difference, but she did not understand.

When we moved into the city, the situation became dangerous because she attacked every dog in sight. I tried the discipline methods I have mentioned, but they failed. The brain pattern to attack was so deeply set that when she saw a dog she automatically attacked. She became so intent on the chase that everything else got tuned out, including my voice calling her. She never even heard me. After the attack she would realize she was wrong and come back, knowing she would be punished. She was so perfect in every way that I had to resort to harsh methods or have her destroyed. I obtained a cattle prod stick, and the next time she went after a dog, I caught her and shocked her with it. At the same time I sent her the image of the attack so that she could visualize it; it was then perfectly clear to her why she was being punished. The treatment was applied twice. The third time she started to attack I called her; she heard me and stopped. Each time she responded favorably, I praised her excessively. Gradually the pattern of reaction changed, and she no longer attacked other dogs. Always remember to reward good behavior and punish bad behavior, but never use food as a reward. Food is just a common part of life, and if used as a reward, it can cause eating problems. If you use food, your animals will only respond when you give it to them instead of responding because they love you and enjoy the reward of praise and petting.

Some pet owners believe that separating two pets which do not get along (such as locking one up in a room) will curb agressive behavior. Quite the reverse is true. Confinement to a room only increases agression and confinement to a cage is even worse. I know of misguided owners who like to keep their small pets (toy dogs and cats) *in cages,* unaware of the terrible physical and emotional results. Animals, no matter how small, should never be caged unless they are traveling. They should have plenty of room for exercise and play. This will reduce agression and the results of frustration, such as nervous scratching and chewing.

No sign of aggression toward the owner or toward the animal's doctor should ever be tolerated. It could cost the animal its life. I worked for a veterinarian and had to watch him tranquilize animals before he could treat them. In the case of an emergency, when time is of the essence, the extra time waiting for the anesthetic to work or the physical effect of the anesthetic itself in a weakened animal could cost it its life. I am appalled at the people who will laugh at the animal who bites the doctor who is trying to treat it, exposing their pet to possible injury when the doctor only wants to help. I just cannot understand this illogical attitude and am constantly amazed at the wonderful doctors who continue to care for these animals under the circumstances. This aggressive behavior is not necessary, as I have proven with my own animals who will not even lift a lip at their veterinarian; they know better than to even try.

Many people wonder why their pet knows when it is near the vet's office. Well, it is because the animal can read your mind, and you are thinking that you are nearly there. You are also anticipating a problem or anticipating its fear and consequent biting, and it does what you expect it to do. It reacts because of your fear; so if you will relax, expect obedience and calmness and no problems, and discipline any aggressive behavior toward the doctor, your pet's visits to the vet can be a relatively pleasant, trouble-free experience for all involved.

Insecurity and Shyness

Dealing with insecurity in a shy dog that fears strangers, noise, or any unfamiliar situation demands forced exposure. I know of no other cure. As I have stated before, they do not like the forcing process, but they tell me how grateful they are to be cured of the fear.

Place the animal in a choke collar attached to a leash so it cannot slip out of the collar and get away. Hold the leash near the dog, and walk toward the person or thing it fears. As you approach, hold the leash firmly and do not allow the animal to back off as much as an inch. At the same time convey by visualization that this will be a happy, friendly event. Hold the animal this way until it stops pulling back and begins to relax. Then lavish it with praise.

Philea, for example, was very afraid of people. When I asked her why, she told me that she was afraid people were going to take her away from me again. I had sold her as a puppy when I discovered her mother, Princessa, and her brother, Loverboy, dominating her to such an extent she was becoming painfully shy. She had believed that she would always live with me. When I got her back again after a year, she would not let anybody near her. I took her, with her mother and brother, on short trips. At first she was so afraid she would be left that she would not leave my side to play, eat, or even urinate. Gradually she gained some confidence but still would not let people touch her. I took her to busy, crowded places and held her on a tight leash, forcing her to bump against people in passing and also forcing her to stand still while people petted her. I took her to boarding kennels and left her for a day or two at a time so that she would trust me when I told her I would be back and she should relax. Gradually she began to believe me, but she still refused to allow people near her when she was off the leash. One day I realized how much she loved to be scratched on the rump; so whenever anybody reached out to her, I asked that person to scratch her rump first. Soon she associated people with this pleasant experience, and when she met someone new, she turned around and waited for her scratch. This may seem funny, but I would much rather she back up to someone in a greeting than run away. Now after two years of socialization she is completely over her shyness. For dog shows she must be friendly, especially in the ring. Also, if a frightened dog is lost, it won't allow anyone to touch it to return it to its owner.

Philea also had to overcome her fear of fireworks and sudden noises. Many pet owners face this dilemma during the holidays. Their dogs become so frightened they may tear through screens and barricades or jump fences and walls and end up hopelessly lost. I am against the use of tranquilizers as a way of helping them cope because the tranquilizer merely dulls their reactions and does nothing to remove the fear. When an animal is afraid, it is very unhappy. This is how I helped Philea get used to noises.

I have found that the animal is afraid because it does not know what is "behind" the noise, that is, the cause. An animal must actually see the gun or fireworks or whatever is causing the noise it fears. I took Philea to Mexico during fiesta. The campgrounds were bursting with noise and fireworks, day and night. At the first sound, Philea took a nosedive for the car floor, closed her eyes, and lay there trembling. No amount of coaxing would bring her up to eat or play. After twenty-four hours I felt so sorry for her that we took a ride into the quiet country to get her to relax. This she did, and as we returned to the campsite, she was feeling quite happy. She was sitting in the back of the stationwagon, looking out at the crowds, when she happened to see the next fireworks shoot off. She had no chance to close her eyes and hide and literally froze until it was over. When nothing else happened, she turned to me and asked, *Is that all that noise is?* She realized the flash of light caused the noise, and that was the end of it; it had nothing to do with her in any way. After facing her fear, she willingly left the car and has never again shown any fear of noise. I have also treated sudden noises in a matter-of-fact manner and joked with her about her flinching. The problem is completely gone now.

I have one word of caution about forcing an animal to face an unpleasant situation. Hold the animal on a tight leash because if it is allowed to back away the fear becomes more deeply ingrained in the brain pattern. Also, be sure the animal does not lean against you because then you become its security blanket. Your pet must stand alone or be held away from you in order to gain self-confidence. I do not know how or why it works this way, but it does.

Disciplining a nervous horse that gets spooked on the trail must be done immediately. First, it must be made to stop until it ceases the nervous prancing. When it is calm, make it proceed along the trail. If you take it back to the barn instead of making it face the fear, the fear becomes more deeply ingrained. The horse then becomes what we call "barn sour," trying to return to the barn all the time. Many horses tell me they really enjoy going out, but it has become such a habit to turn and go home that they do it without knowing why. For this reason you should never allow a horse to gallop or canter home but make the return journey a relaxing exercise.

Other Bad Habits

There is nothing more irritating than a constantly barking dog. Chronic barking for no apparent reason is a badly ingrained habit which the dog cannot help and which you must stop. The habit often starts as an attention-getter; then it becomes neurotic behavior. Animals tell me how miserable they feel when the habit takes over and how sore their throats become.

To correct this problem, first try the milder forms of discipline—scolding, spanking with the belt, or restraint. If this does not work, you will probably have to resort to the bark or sonar collar (Cost: $15.00 to $30.00). This is a battery-operated device hooked to a collar; it is guaranteed to break the dog's barking habit within three days. The dog's bark activates the collar to produce a light shock. This shock breaks the brain pattern in the same way as the hot-shot stick. The dog quickly learns that barking *hurts;* so it stops barking. For defense, a dog can protect itself by growling and biting. This is still possible with the collar because only the high pitch of the bark will activate the shock. Be sure, however, that there are no other barking dogs nearby because *their* barking can also activate your dog's collar. I knew someone who owned a Samoyed dog and took it to a dog show. The owner had to leave the dog and did not want him barking; so he put the collar on the dog and left. The poor dog went into a state of shock within a few minutes because all the other dogs around him were barking and activating his collar. The poor animal didn't know why the shocks kept coming. Those of us nearby cut the collar off him and calmed him down until the owner returned. You can imagine the reprimand awaiting that person! After your animal is trained, you can get a collar that *looks* like the sonar collar but without the batteries. This is usually sufficient to keep your pet under control.

Car chasing is a danger to all concerned: your dog, the child who may run out after the dog, the driver who may brake suddenly to avoid hitting your dog or veer suddenly and crash into another vehicle, causing an accident and possibly deaths for which you are legally liable. For everybody's safety, your dog must not be allowed to chase cars.

Training can be accomplished by a variety of methods. One is the squirt gun, containing plain water or a combination of water and vinegar (diluted in a ratio of three-fourths vinegar to one-fourth water) to insure that the mixture is not dangerous to the animal's eyes. A liquid soap bottle (carefully cleaned of soap) is a larger and more effective container. Arrange a ride in a friend's car. As your dog begins to chase, squirt it directly in the face. Two or three of these squirt-gun treatments should break it of this dangerous habit. If not, then the shock treatment is in order. You can use the hot-shot stick, which I used to break Princessa from attacking other dogs, or you may use a remote control shock collar. This is a device similar to the bark or sonar collar, but you control the shocking device. Watch the dog from a hidden vantage point, and when he starts after the car, activate the collar to give him a good shock. This will break him of chasing cars, jumping fences when you are not present (or so he thinks), attacking other animals, or other behavior that must be curbed. The remote control collar is rather expensive (about $200.00), but you may be able to rent one from a person who trains retrievers. It is used to shock a dog at a distance when the dog bites down on a bird it is retrieving or to get its attention at a distance when the dog refuses to respond to the voice or whistle command.

The squirt gun with plain water is also the most effective discipline instrument for cats that insist on fighting. It will also stop a spraying job in process.

Group Discipline

Group discipline is a very interesting concept. Mary Roberts of Covina, California, owner of German Shepherd champion Gauss Wikingbluet, clued me into this unusual practice. Mary's first two Shepherd studs were trained not to leave the property, but not Gauss, who was new to the family. The first time they were in the front yard Gauss promptly ran off to investigate the neighborhood. The other two Shepherds followed Gauss to the edge of the property and waited. When Gauss returned, the other two dogs were so angry that he had been allowed to leave while they were not (they explained this to me when I questioned them) that they attacked him, snarling and biting. This, Gauss told me, made him stay at home. He had no intention of getting beaten up again.

I decided to use this theory of group discipline on a problem I had with Loverboy. Lover sometimes barked incessantly for attention and sometimes just for the fun of it. When I explained to him that he'd have to stop this barking because he disturbed the neighbors and me, he continued making a racket. I tried the standard forms of discipline previously discussed, except for the shock devices, without success.

It was definitely time for a chat. I sat down with Lover, Princessa, and Philea and told them the next time one of them began barking needlessly I would whack them all. Later, Lover began to bark. First, I made sure there was no reason to bark. Then I used the plastic belt on all three dogs. As soon as I had finished, I explained by mental communication why I had spanked them all. Later that evening Lover barked again. At that point Princessa promptly attacked him and put him in his place. He has since learned to bark only when something is present. I make a point of praising him for that necessary barking, as well as disciplining him for the foolish unnecessary barking. Group discipline is a fascinating parallel to what we in human circles term peer pressure.

All these training methods must be administered with love and concern for the good of your pet, as well as respect for your neighbors' rights to privacy and quiet.

10

FRUSTRATIONS AND FUN

This chapter also deals with discipline but on a more sophisticated level. This level involves showing your pet, a consideration of obedience classes and what to look out for in them, and attack training for guard duty. I also discuss neutering and spaying here because it is not only basic to an animal's personality and health but also, in many cases, its ability to obey commands without distraction. Finally, because it is such a rewarding experience for my pets and me, I suggest rules for traveling together. Discipline and good manners are even more essential on the road.

Showing Your Dog

You do not have to be a professional breeder to have a good animal and want to show it or to want to show the offspring of your purebred bitch.

First, a warning. Don't take your pups to puppy matches. One or two or even three matches will condition them to the show situation. More than three matches will make the dog so bored that by the time it is entered in serious competition it will have lost that animation which is so important if the dog is to show well. If you, the owner or handler, need practice in showing your animal and want to attend handling classes and matches, borrow another dog for the purpose. Or take one of your own dogs you don't intend to show. The object here is to give *you* the practice, not the dog. The dog remembers quickly and will perform as you have trained it. Practice bores them, they tell me, and so they misbehave.

Second, remember your visualization. Don't be obsessed with your nervousness and the negative picture of the dog running out of the ring. Be positive and confident, and your dog will get the message loud and clear.

Third, keep an ace up your sleeve. Discover something special that your dog loves to eat or do. Do not give it this special food or activity until you have been showing your dog for some time and have noticed a lack of interest or animation. Then start giving the dog that special treat or activity right after it comes out of the show ring. This will then be connected with the show ring and will reactivate its interest. Many times a dog gets nearly all its points toward championship, but by the last few shows it is bored and disinterested. Also, the owner or handler feels a great deal of tension because these last shows are so important. This makes the dog nervous because it is able to feel your tension. The treat or special activity helps ease that tension, but you must also concentrate on relaxing in order to help your animal do a better job.

A final tip: If you have a male with strong hormones and you wish to show him, you may have problems. He will become distracted in the show ring when he smells the presence of females, especially a bitch in heat. Ask your veterinarian for some female hormones to give the dog about half an hour before he enters the ring. This will make him feel as though he has had the mating experience and will relax him. These hormones do not cause cancer, another old wives' tale, or create problems later if and when you wish to breed him.

Obedience Classes

Many people attend obedience classes, hoping to help their pets and relate to them more effectively. Most obedience classes do just this. They help owners develop a consistent, happy relationship with their pets. A good class can be defined or measured as teaching obedience through kindness. Nothing is done to hurt the animal, and nothing more than a jerk on the choke collar is allowed to get the dog's attention. Training should be fun for you and your animal. Unfortunately, any number of licensed obedience classes use brutal methods.

Here's what to look for and avoid. Notice the animals before you enroll your pet. If you see an animal in a class or in an obedience trial that walks with its tail between its legs, head drooping, or body trembling, this is not the place to bring your pet. Subdued behavior means an animal has been bullied into obedience, not educated. Observe the performances, and take note of the dog that wags its tail, runs to obey the command with a bounce to its gait and with its head up. Try to find out where that owner trained the dog, and attend those classes.

Two examples of authorized animal abuse took place in a free obedience training class which I had hoped to attend with Princessa. What went on during that one evening I would not have believed unless I had seen it with my own eyes. The trainer was using the Bill Koehler method, which is geared to training guard and attack dogs. Or rather I should say the trainer was abusing the Koehler method and adding some innovations of his own. A young ten-pound Schnauzer was being taught to heel. I watched the trainer take the dog on a six-foot leash and choke collar. When you are teaching the heel command, the dog is supposed to move forward at the word *heel.* As you move your left foot forward, it must stay by the left side until that foot comes to a stop at which time the dog is to stop and promptly sit. As this trainer said heel, he stepped forward with his right foot; this delayed the dog's start since it was following the left foot movement, which was correct. When the dog did not move immediately, the trainer gave the leash such a hard jerk that the dog was flung forward like a whip, flipped over on its back, and landed on its back on the cement. Strong jerking on a choker chain can damage the cartilage in the trachea, resulting in severe breathing problems which often require surgery. After that, the dog was so frightened that when it heard the command *heel,* it shot forward to keep from being flipped again. The dog was not being educated but bullied.

The second example involved an Irish Setter that was being used to teach the class (about forty people) how to break a dog from jumping up on people. The trainer was told, in the presence of everybody watching the class, that the dog had been spayed the day before and was still very sore. Nevertheless, this trainer brought the dog out into the ring, and said, "Up, up!" pointing to his chest. The dog jumped up. The trainer grabbed the animal's front paws, held them, and kicked the dog in the stomach with his knees. This action sent her reeling backward. She too landed on her back on the cement. She staggered to her feet, in pain and confusion: She had just been horribly punished for obeying. What she learned was fear of commands, not to stop jumping on people.

I also observed this particular trainer using the hot-shot stick for the slightest infraction of any obedience rule. This is against everything I believe about animal training, and the only purpose in reporting these incidents is to warn you what to look out for, especially the use of electrical discipline devices in the hands of a sadistic person. Animals feel pain just as humans do.

Attack training is used in training police, military, and guard dogs to attack and release a person on command. I am against the most common form of training. Here a person wrapped in heavy protective clothing aggravates the dog until the animal is angry enough to attack the person, at which time it is also *told* to attack. However, there is a new school of training that is getting the same results without aggravating the dogs. The trainers make a game of attacking and then praise the animal for attacking when told. I cannot give you all the details here, but I suggest you investigate this particular point when training your dog for guard. The animal should protect and attack because it wants to obey and please you, the owner, not because it hates people.

The methods I have described are often used by established trainers in legal or licensed classes. You will want to protect your pet. I urge you to use mental language as well. In other words, use visualization of the positions commanded and transmitting of feelings as a part of the training. Expect more from your animal and give it more respect. The response will be phenomenal.

Spaying and Neutering

Unless you are going to breed your animal, I strongly recommend having it spayed or neutered. Apart from the animal population explosion and its tragic results, which I will discuss presently, I recommend spaying and neutering for the animal's well-being and health.

According to the animals themselves, the female dog or cat is miserable when she is in heat; she experiences the sex act under duress and without affection or any knowledge of the consequences; she becomes tired, listless, and physically ill with constant pregnancies; she suffers while giving birth. She feels helpless and may not even want her litter. I have known cases in which the female did not want the litter to begin with, and so she reabsorbed the fetuses while still in the womb. Other females have killed their newborn offspring—by refusing to nurse them, pushing them out of the basket, or by lying on them and smothering them. Female rabbits and mice have also eaten their young. Merely bearing the litter does not insure mother love.

Another problem for your unspayed female who gives birth is arousing the maternal instinct. Many people believe the old wives' tale of having the animal give birth to one litter and then spaying it. This contributes to the millions of unwanted dogs and cats. It can also affect the emotional well-being of the spayed animal who may then become frustrated and try to "mother" every small animal she can find. It is a heart-rending experience to watch.

It is much healthier for your female pet to be spayed before her first heat cycle. A female cat should be spayed at six months; a female dog also at six months.

The male dog or cat fares somewhat better in the sexual arena, but he is also helpless. His hormonal make-up causes him to mount any female in heat, regardless of how strong he may be. It's pure drive. Sex in animals is for reproduction, and reproduction is no more essential for them than it is for humans who have the ability to choose their partners and whether or not to have families; yet sex and reproduction have distinct results on an animal's health.

Neutering the male dog or cat, as mentioned before, will curb overaggression. According to many veterinarians, neutering the male dog can reduce or nearly eliminate anal gland problems, a difficulty often chronic in smaller breeds of dogs.

Neutering cats has an additional advantage: a longer life span. Veterinarians have told me they have never seen an unneutered male cat "grow old," an astounding observation. The normal life span of a cat is nineteen to twenty years, but this is only for spayed or neutered cats. Unneutered males frequently get into fights over females in heat which result in serious injuries, infections, or even death. Infections and abscesses can cause kidneys, liver, or other body organs to break down. Many animals die from these ailments, but the deaths are not connected by the owners with the lack of neutering or spaying or with the violence incurred by the sex drive in the animals.

Spaying and neutering of cats will do much to curb the dangerous problem of overpopulation and its tragic consequences. It has been estimated that there now may be over twenty-five million homeless cats in the United States. Many parents allow their cats to breed so that their children may witness the act of birth. This is gross irresponsibility. Children can receive education about animal reproduction in school, without adding to the animal population.

Contrary to common belief, spaying and neutering does not make your animal fat. Only food does that. After surgery the animal requires food, but not as much as before. My advice is to cut down on the amount of food you give your pet until the animal's hormones have had a chance to balance out again. Then normal weight will return.

A look at canine mammary cancer development and postsurgical survival is another good reason for spaying a female pet. Spaying reduces the chances of an animal's developing mammary (milk gland) cancer by nearly 200 percent. According to the *Journal of the National Cancer Institute* (December 1969), mammary cancer was found to be the most common malignant neoplasm in the female. At that time, the incidence rate for mammary cancer accounted for nearly one-half the deaths of older unspayed female dogs—either from the cancer itself, or through death from surgery to remove the cancer in the mammary glands.

There are two more reasons for neutering and spaying. One concerns the problem of what to do with unwanted animals. The other is legal and financial.

You may, for one reason or another, have a pet or a whole accidental litter you cannot care for and/or do not want. There are three ways of getting rid of these animals. One way is humane. Two are needlessly cruel. Adoption is a fourth, a separate category.

The humane way is to take your animal to the veterinarian.

The vet will give your pet an injection, and it will just go to sleep, quickly and painlessly.

One needlessly cruel way is to take your pet to the local animal pound. Here your pet is shoved into a cage filled with other animals, some of them fighting, all of them experiencing rejection, shock, loneliness, pain, and terror. It is bedlam or hell. When a sufficient "quota" of dogs and cats is reached, the animals are dumped into a large dark chamber. The door is clamped shut, and the air is siphoned from the chamber so that all oxygen is drawn off or expelled and the animals cannot breathe. They die an excruciating, suffocating death. Their asphyxiation is made all the more agonizing by the biological process involved. The nitrogen in the animals' blood boils so that they also suffer an agony similar to that of a deep sea diver with the bends. The victims then stay in this chamber for several hours before being shipped off as one of the raw materials to be made into fertilizer. Despite the efficiency of this system, some animals do not die from asphyxiation; they lie there in the dark, in pain and fear, beneath a pile of dead bodies. My research has revealed that the majority of animal pounds in this country destroys animals this way and not with the needle, the painless injection. Why? Pet Finders, Incorporated, in Los Angeles did a study on the difference in cost between the chamber and the needle. I spoke to Jody King, its president, who was involved in this study, and she said it was actually *cheaper* to destroy animals by injection. But since the chamber method is already in existence, the pounds are reluctant to change.

Putting your pet up for adoption seems humane, but the moment you place an animal for adoption, you legally relinquish all rights to that pet. It is then disposed of at the discretion of the animal pounds. Publicly supported animal pounds are *not* obligated to find a home for your pet, and the percentage of animals adopted from pounds is minuscule compared to those killed in the chamber or sold for laboratory research purposes. By "research," I mean vivisection and various barbaric experiments perpertrated on animals in the name of science.

I remember observing a public school science class, where the teacher was instructing about thirty students in the disection of a live turtle, to expose the functioning organs for observation. Each student had his or her own turtle. The cruelty horrified me and it must have affected the children. They had already witnessed a film of actual surgeries, so that these particular lab experiments were totally unnecessary in terms of education, and could only have been a lesson in cruelty. I do not believe this public school is an isolated example. I believe the practice is widespread and its effect utterly damaging to humans and animals alike.

Another cruel way of getting rid of your pet is to "dump" it. The domesticated animal is unable to hunt for food; it must be taught this by its mother or another animal. When you dump a pet in the country, it does not know how to forage or kill for food, and it will starve to death. Dumped pets also don't know how to defend themselves from predators, any larger animal that eats a smaller animal, such as coyotes. Farmers, who cannot afford to feed these strays and certainly don't want them after their livestock, either shoot them or poison them. Animals dumped in the city fare no better. They may also starve to death, get poisoned, get hit by a car and suffer a slow death, unaided; they may get injured in a fight, die of disease (because their immunization shots have worn off), or get caught by someone who will turn them over to the local animal pound to die the death I have just described.

Adoption, the fourth category, means finding a home for your pet *which you yourself know will be a good home.*

If you are able, both psychologically and physically, you may put your pet to sleep just as painlessly as the veterinarian does. Newborn puppies and kittens may be painlessly chloroformed or drowned in warm water minutes after birth. But you must be aware of the mother animal's condition and take her to the vet. She will be lactating, and her entire litter should not be taken from her. She should have one or two of the litter so that her milk will dry up slowly and emotional trauma will be avoided.

A final word on the subject of neutering and spaying. The advantage discussed here is technical: Who is legally and financially responsible for the veterinarian's fees for abortion or for the cost of the loss of the litter in the case of a registered female for which the owner has already paid a stud fee? It's all dollars and cents, but it can add up—if, for instance, your male animal gets to somebody else's female on her property, or vice versa. I shall use my dogs as an example.

First, a bitch can have puppies from each male breeding her during the same heat cycle, and nobody can possibly tell which puppy is sired by which dog.

I pay a two-hundred-dollar stud fee. If my bitch is impregnated again by another dog, there can be no proof that the puppies are the offspring of the registered stud because nobody can determine which ones are his. Since it was not the owner of the stud dog's fault that my bitch got impregnated by another dog, I cannot reclaim the two-hundred-dollar stud fee. That is lost. The litter is considered a mongrel litter, or unwanted puppies. So it must be aborted—or I'll have unwanted puppies. A litter can be aborted in one of two ways. The veterinarian can give the mother a hormone shot within five to seven days after mating followed by another in seven days. Or the dog can be spayed while she is in her pregnancy. In either case I am responsible for vet bills which, in the case of spaying a German Shepherd, can run as high as two hundred dollars, depending on the veterinarian and the area of the country.

There is a further financial loss to be considered. My German Shepherd puppies sell from one hundred fifty to three hundred dollars or more, depending on the show quality of the pups. Shepherds have anywhere from six to eleven puppies in a litter. This means a loss of about fifteen hundred dollars in sales. Since I cannot breed the female again that season, my loss encompasses the whole year. In the case of a bitch that has produced champions, her pups may sell for as much as eight thousand dollars. If there is a bitch around, it obviously pays to neuter your male or make sure he does not get out and mate her.

Costs vary from region to region. Currently the range for having a cat spayed is twenty-five to fifty dollars, and for neutering it is fifteen to twenty-five dollars. Spaying a dog depends on the size, and the range is from twenty dollars to sixty-five dollars. It is slightly less for neutering. Another financial reason for spaying and neutering is the high cost of handling complications in delivery. Caesarians are more common than is realized and can cost up to two hundred dollars. With a mongrel litter you cannot cover the cost in sales.

Travel

Many people with pets cannot get away because they don't know how to travel with their pets. Boarding them is expensive, and some animals definitely don't like being in kennels. One way or another there is frustration for either owner or pet. This need not be. A psychologically sound animal can travel anywhere. I travel all over the country with my animals, a companionship we all enjoy, and a practice that means safety for me. If you feel like bringing your pet along on your trip, here are some guidelines. I am referring to road travel—by car, trailer, or camper—and to dogs and cats. These animals can be delightful companions as long as you meet their requirements on the road with the same respect and consideration you give yourself.

An animal's main requirement is exercise. On a long trip, plan to set aside one hour each day to let your animals run about. The best form of exercise for a dog is to teach it to run and fetch. Try to find an open field or some other isolated area. A lake is even better. Swimming is excellent exercise for a dog.

I fling a floating weight out over the water. The dogs love fetching it, and soon they are relaxed after those hours in the car. Dogs have a thick undercoat so that even very cold water is enjoyable. After their swim, dry them off with old towels. If your pet is a working breed (Shepherd, Samoyed), you have additional options. My Shepherds wear harnesses on occasion, and I have taught them to pull carts. They love this activity. They also get exercise pulling children on skateboards. All fun and games for an hour. When do we get on with the trip? My dogs are trained to return when I blow a whistle. Any tardiness is punished by restrictions. They soon learn.

Cats can also be good travelers. Just be sure to keep a litter pan handy at all times. Don't be upset if your cat seems nervous or yowls a great deal during the first few car rides. Most cats are transported only for a visit to the veterinarian. Their memories of car travel are understandably unpleasant. But you will find this nervousness passes as the cat begins to view the car as a place of security.

When I first began traveling with my cats, Snow Bunny and Tigger, I kept them on leashes. A few times they escaped the camper before I was able to collar them, but they soon returned. At night the cats were restless and begged to go out. I communicated with them and said I would respect their need for exercise and allow them out at night, provided they returned first thing in the morning at the sound of the whistle. I also began feeding them in the morning; so they were also hungry then. During the night I keep the side window of the camper open for easy entry. If for any reason the cats do not return when I blow the whistle, or by the time the dogs come back from their morning run, the cats are disciplined for their tardiness and not permitted to go outside the following evening. It is a workable arrangement. Each morning the dogs are let out to play in the fields, but always within my sight, and about half an hour later they return in response to the whistle, with Snow Bunny and Tigger prancing behind. I always park near a safe, open area. Never let dogs or cats run free in a suburban or city area.

If your pet is old, it will require extra consideration. At fourteen, Blacky had certain physical disabilities; so I always brought along his basket and bedding. These familiar comforts give him a sense of security.

I consider one safety precaution essential. When traveling, be sure that all your pets are fitted with identification barrels. These barrels are sold at a nominal price and can be bought in most pet shops. The use of barrels, rather than tags, makes it possible to change addresses easily. Type all the necessary information on a small piece of paper and place it in the barrel, which locks tightly and is also waterproof.

If your travels include motel or hotel stops, be prepared for some opposition from the management. Sometimes it is possible to persuade the manager that you understand his or her position and offer a list of rules you intend to follow. *Then follow them!* This will persuade the management that pet owners and pets are not necessarily destructive to property and the comfort of other guests.

Rule One. Never leave your dogs in the motel room alone while you are gone. Strange noises, people constantly walking in the halls, and the unfamiliarity of the surroundings may frighten them or cause them to bark or chew furniture. Leave them in your car; this is secure familiar territory. But be sure to leave a window slightly open, for in hot weather animals can die of heat prostration.

Rule Two. Always carry a disposable bag and a scooper of some kind. Pet owners who allow their dogs to defecate on motel property cause considerable resentment for obvious reasons. This hygienic act should be applied to public parks and beaches.

Rule Three. This may not be applicable in all cases, but I personally feel that if you want your pet to travel with you it must be spayed or neutered. The spayed female will not attract male dogs (or cats); nor will she leave blood stains on motel carpets or furniture. And when the next tenant enters with a nonneutered male, this animal will not get a sexual scent which will excite him to spray on bed or curtains. *Any* scent of another dog lingering in the room creates excitement and pressure on the prostate gland of the male, and he will lift his leg on the furniture. A dog will also wet to mark its territory if it has not been neutered. Is it any wonder then that motel owners refuse to allow animals in their rooms or at least need some considerable persuading that you and your pet are respectable guests.

Another firm rule for a motel stopover is to keep your animal on a leash. It will not be able to run "off limits" and into flowerbeds or get into other trouble with other guests or other guests' pets or the management.

Well-adjusted pets can be enjoyable traveling companions and can add new dimensions to crosscountry trips, either business or pleasure. My animals travel with me, cats and dogs, and it is an adventure for all of us, particularly me, because they see things so differently from me, and they share their adventures with me.

When you are traveling with your pet be sure that its vaccinations are fairly recent and that you have a health certificate issued by your veterinarian indicating that the animal is free of any communicable disease to present to health authorites if you are asked. This health certificate is a requirement for traveling on public conveyance—trains, planes, etc.—as well as for interstate shipment or transport outside the United States.

Traveling abroad with animals will necessitate very different adjustments for all concerned. Check the quarantine laws for the country you want to visit. England demands a six-month kennel stay for quarantine purposes. Upon reentry into the United States, there is a three-month kennel stay. You can visit your pet in quarantine at specified times, but the relationship that may develop over this period of time and under these particular circumstances may be frustrating and generally undesirable for your pet. The physical restraint, loneliness, and so on will usually cause trouble for the pet, physically or psychologically.

If you are going to Mexico, I recommend leaving your pet behind, either at home or in a familiar kennel. In Mexico a great many stray dogs wander around; they have not had proper innoculations and are scavengers. They are starved, very possibly diseased. If your pet is found by Mexican authorities among the strays, it will undoubtedly be destroyed. Not all travel plans can include your pet.

The guidelines I've suggested in this chapter have been worked out by and between my animals and me, and they tell me they love to travel with me. It is a growing experience for all of us, a new adventure for them and a new experience for me just watching their enjoyment. I see the world as they see it, and this is a revelation. I also love their company during the long, lonely hours of driving.

11

CONSUMER AWARENESS: PETS AND PET PRODUCTS

The pet industry represents a multimillion dollar business whose promoters sell their products through advertising that will appeal to pet owners. Like many other enormous industries, the ad campaign does not necessarily assure the facts. A consumer is wise to know the contents of a product. Reading the label is easy. Understanding it may be more difficult. Advertising claims may be deceptive; so here are a few guidelines.

Pet food ads, for example, are designed to appeal to the *human* appetite, and *appearance* is important. But animals are animals, and my animals clearly do not care what the food looks like. They are also unaware of its nutritional value and whether it is good for them. Taste and smell, however, are important. Television commercials do little to explain the contents of a product. They show animals eating hungrily. Possibly there is meat buried at the bottom of the bowl, which the viewer cannot see. Possibly the animal has been starved for quite some time before the commercial is made so that the animal will really eat with gusto. I know these possibilities exist because I have worked in television, and I know some of the trainers responsible for animal commercials and the tricks used. They are not abusing the animals, but they are making certain they get the proper reaction during the filming. Many animals are trained to eat on command. True, sometimes the animal really gobbles up the food, but that does not mean it is a healthy diet. Animals do not have the ability to choose what is healthy for them or know their dietary needs any more than children do. But they do go for taste, and this is what the pet food manufacturers know: Taste *sells.*

Food

Most animals, dogs and cats alike, love moist packaged foods. This is because the sugar content may be as much as 33 percent. Most moist packaged foods represent the same lack of nutrition as a bag of candy. They are also high in salt, a preservative. According to some veterinary authorities I've consulted, this combination of high sugar content and salt could be the cause of the recent increase in the incidence of kidney problems. These vets have told me that it is similar to a human being eating high amounts of sugar or living on sugar-cured ham. In my opinion it is best to eliminate or nearly eliminate moist package foods from your animal's diet.

I am against the use of canned foods except as a rare treat. They contain large amounts of preservatives and food coloring, as well as waste products. Recent studies on the effect of food dyes on humans make one wonder if food dyes do not also affect animals. I have found that dogs fed canned food or the rich packaged foods seem to develop "doggie" odor more than dogs fed on the type of diet I recommend.

When the label says "byproducts," it refers to lungs, intestines, and other entrails. True, it is natural for a dog to eat meat, but an all-meat diet is unbalanced. Meat is high in phosphorus, and the animal needs a calcium supplement because a calcium-phosphorus balance must be maintained in the body. Lacking calcium in its food, the animal will pull calcium from its own bones. I once saw a cat that had been fed just meat and liver for several years; eventually its bones became so porous and brittle that it could no longer walk.

When the label says that there are "grains" present, that is good for your pet. The animal needs a balance of the carbohydrates and minerals derived from grains. In the wild, animals eat fruit, grains, grasses. And when they have killed to eat, they have eaten the bones and skins. Read the label carefully to be sure of a balanced diet of (1) meat or protein, (2) grains, (3) oils for lubrication and better coats, and (4) minerals.

There are special diets for special problems, but I do not recommend giving them to your pet unless it has that specific problem and it has been so diagnosed by your veterinarian. I do not recommend using RD (Reducing Diet), ID (Intestinal Diet), KD (Kidney Diet) or any of the other special diets unless it is prescribed for a particular problem and is advisable.

Dry foods are good for animals because they stimulate gums and help keep the teeth clean. However, a small amount of *raw* meat (cooking meat destroys the enzymes in it and makes it harder for the animal to digest) or eggs or cheese must be included in the dry food diet to balance the carbohydrates. Dry food alone, because of the lack of oils and digestible protein, results in an excessively dry skin and dull coat. So add a little cod liver oil or cooking oil to the food, along with the protein supplement. Most commercial dry foods do contain a protein that meets government standards, but it is keratin. Keratin is a dead protein that cannot be assimilated. In my opinion and in the opinion of many veterinarians, the best and most complete dry dog food on the market at this time is Science Diet, made by the same company that produces the special diets for veterinarians. It can be purchased through vets' offices and pet shops. It is not on the usual supermarket shelf.

My cats thrive on dry food and one or two ounces of raw beef kidney. There is a great deal of controversy over the use of dry cat food because some veterinarians think that the high ash content causes kidney problems. Ash is a waste product with no nutritional value, and it is used by pet food manufacturers as a filler; it creates bulk which is deceptive to both consumer and cat. Most cats do not drink enough water to flush out the kidneys and eliminate the ash. The ash then becomes concentrated in the kidneys and may cause urinary problems. Some cats and dogs are allergic to milk, which may give them diarrhea. But if it does not, milk is a good source of calcium for both cats and dogs and is a way to get them to increase their fluid intake and flush out their kidneys. Whenever a cat or dog has a kidney problem, meat should be eliminated from the diet. The metabolization of meat in either animal produces ash as a byproduct (or waste product) and the weakened kidney cannot handle it, causing further damage. Meat is not an absolute essential. I have seen animals with diseased kidneys live for years on a meatless diet and plenty of fluids. I have also found that one or two ounces of raw beef kidney each day will aid in preventing cystitis in cats. One of my cats, Snow Bunny, had chronic cystitis. One day when I decided to take her off her medication, blood immediately appeared in her urine. She now gets raw beef kidney every day and has never had a recurrence. A Mrs. Henderson of Sacramento, California, had the same success curing her dog's cystitis. Do not cook the beef kidney. Something is destroyed in cooking that helps the animal's kidney. I once helped a cat whose condition was so toxic her body was swollen and her appetite gone. The owner blended the raw kidney with a little water and forced it down the cat's throat with an eyedropper. Within twenty-four hours the swelling had disappeared, and the cat's appetite was normal. In addition to dry food and raw beef kidney, my cats also get raw fish and other raw meat from time to time. Cats enjoy variety, but the greater variety you feed a dog the fussier it gets.

A word about cooking and not cooking poultry. Chicken necks and backs are cheap, and they are also of nutritional value if given raw. Raw, they provide oils, calcium, and protein. Never cook chicken bones for your pet. Steak and chop bones are equally or even more dangerous; they become brittle and break in the throat. Laceration will occur, and the animal can choke to death.

Food Supplements

There are various food supplements which I must mention here. These include Brewer's Yeast as a total B vitamin source, alfalfa, specific vitamins for specific needs (health, illness, pregnancy, old age, and newborn), and other sources of minerals. When I speak of vitamin supplements, these are the same vitamins I take myself, purchased at health food stores where vitamins are made from natural sources. Synthetic vitamins are a petroleum byproduct.

Alfalfa and Brewer's Yeast produce healthy coats, minimize odors, and aid or prevent skin disorders—along with a well-balanced diet. Alfalfa contains a natural chlorophyl; it is also a complete source of minerals. Cats don't like the taste of alfalfa; so I disguise it in their food. On the other hand, most cats love Brewer's Yeast; so I give them the tablets to chew as treats. Brewer's Yeast helps keep flies, fleas, and mosquitoes from biting your pet. To determine the dosage, try one tablet (7½ or 10 grains) a day for eighteen days. If your pet is still continually bitten, increase the daily amount until you notice an improvement. But do not expect Brewer's Yeast to eliminate fleas entirely. It helps the pet's tolerance because the fleas do not bite as much. It does stop flies and mosquitoes completely. (More about fleas and protecting your pet from them later in this chapter.)

Alfalfa supplement also greatly eases the discomforts of arthritis—with the exception of rheumatoid arthritis. Alfalfa with Vitamin E helps large-breed dogs which may suffer from general arthritic pain, hip problems, and hip displasia. Hip displasia is a major problem with all large breeds, and there is considerable controversy about it. Some people think it is inherited, and others think it is caused by poor nutrition. I believe it is the latter, and I am now engaged in a breeding program to try to prove this. I think it is a lack of Vitamin C (which holds calcium in the bones) and E (which promotes growth) in the diet of the newborn pup until it is one year old, as well as a lack of calcium *or* an inability to assimilate calcium. (I also think pups must be given plenty of exercise to build up the muscles that hold the hip sockets firmly in place.) Dr. Kurtz, our vet, also advises a Vitamin C supplement to treat osteomylitis and other bone problems that hit the shoulders of the larger breeds, especially the Great Dane.

For chronic eye problems, glaucoma, and cataracts, I have had a great deal of success with a combination of Vitamins E and A. Vitamin E helps circulation and gets the nutrition into the bloodstream and the extremities of the body. Vitamin E builds stamina and is given to racing animals. Vitamin A is used in the Rhodopsin series to rebuild the eyes. My Pomeranian, Blacky, had been diagnosed by two veterinarians as having beginning (incipient?) glaucoma. I gave him 100 units of Vitamin E and 25,000 units of Vitamin A daily for about two months, and the condition cleared up. When I saw him getting cataracts, I repeated the treatment, and this seemed to stop the progression.

For heart problems in an old animal I recommend Vitamin E and potassium. Blacky has had a bad heart for about eight years. I give him additional Vitamin E, along with Digoxin, and it is helping prolong his life. After a recent check-up our vet told me that Blacky is healthier than he had ever seen him. Dr. Kurtz has been our vet for three years now.

Since I am not a veterinarian, I cannot, officially, prescribe medicines, dosages, or even the use of vitamins. I am merely telling you what I have found helpful to my own animals—their care in sickness as well as health, their physical needs and emotional needs, and their dietary requirements and how these can best be supplied.

Weight is an important factor in an animal's health. A thin, underweight pet will look starved, and you will act accordingly.

Obesity is less likely to be observed (just chubby like a baby), and yet it can be very unhealthy. An immediate solution: Cut down on the food intake regardless of food package "directions." Different animals metabolize differently. An obese animal can be helped by an RD diet food (Reducing Diet), recommended by your vet. This must be adhered to strictly, with no between-meal snacks. Exercise should be increased gradually at the same time. As with humans, excess weight taxes the heart and can cause early death.

Another reminder about the care and feeding of your pet: Try to consult a veterinarian who is qualified to recommend a specific diet, one who has been trained in nutrition. Most vets have not had sufficient training in this field and are not qualified as specialists in animal nutrition. My recommendations and advice are based on my studies in chemistry, animal physiology, and animal nutrition, on personal experience with my large animal family, and on the nutritional experiments other animal breeders have shared with me.

I am strongly against the use of cortisone because it suppresses the adrenal glands and causes deterioration of the bone. For example, some horses have been given so much cortisone that they have broken their legs, just running, because their bones have become so brittle. Cortisone can control some symptoms, but it does not cure the problems causing the symptoms. The only time I use cortisone is when an animal of mine gets a skin rash and needs immediate relief from the itching, but I also get rid of the rash as quickly as possible.

Fleas, Fungi and Other Problems

Flea infestation is another animal problem, and considerable controversy rages over its prevention and cure. One popular device is the flea collar. This collar can be dangerous to your pet's health and may even prove fatal.

Fortunately several veterinarians are beginning to discourage the use of flea collars that contain organic phosphates. In a paper to the Veterinarians' Association, Dr. Fredrick E. Sherman of La Pesa, California, delivered this message:

> Flea collar toxicity is not an unusual finding in the practice of small animal medicine. Not only the dog, but also the cat has shared in the toxic effect of wearing flea collars. Symptoms of hair loss around the neck with or without damage to the skin are commonly seen in all ages of pets. Most of the hair loss and skin damage will appear directly beneath the flea collar; however, in more severe toxic conditions, the area of skin damage may increase several times the width of the flea collar.
>
> Other areas of flea collar toxicity may be exhibited by increased nervousness, particularly in cats. Major organs of the body, such as the liver and kidney, are overworked due to the body absorbing toxins from the flea collar. The liver has the capacity to repair its own damaged cells, which, unlike any other organ, can sustain a long period of a low daily intake of a toxic material.
>
> In my opinion, this is why some animals can wear a flea collar for several years without showing any clinical symptoms. Unfortunately, the kidney does not have the ability to replace permanently damaged filters, so when the kidneys have endured a severe toxic blow they may be permanently damaged. Not all dogs and cats will respond unfavorably to wearing flea collars, but I do feel that a potential health hazard does exist when any dog or cat wears a flea collar day after day.

Despite opposition, I agree with Dr. Sherman, and so do many veterinarians with whom I work. Many vets will not treat an animal wearing a flea collar because they feel they are treating the toxicity caused by the collar instead of an illness. This is how the flea collar works and why we do not recommend it. The flea collar releases chemicals (in the type I'm discussing, organic phosphates) which are absorbed into the animal's bloodstream, through the skin of the neck. This circulates through its body. As the flea bites the animal's skin, it ingests the chemicals in a sufficient amount to kill it. The chemical is supposed to be strong enough to kill a flea but not strong enough to hurt the pet. I believe that the chemicals affect the pet's nervous system as well as its kidneys and liver. Moreover, the breakdown of liver cells is so gradual it is not observed until the damage is done. Other unhealthy side-effects of this type of collar which I have personally observed include seizures (often diagnosed as epilepsy), lethargy, and sudden irritability in pets that are normally well tempered and happy. The toxic effect depends on the individual animal. Some recover within a few days after the flea collar is removed; others may be dead within those same few days.

A Los Angeles pet owner, Charlotte Currivan, has a Poodle that had been treated for epilepsy for several years by Dr. Kurtz. When I met the dog, I suggested removing the collar for awhile since I'd already noticed a greater sensitivity to flea collar toxicity in Poodles (or Poodle-mixed breeds) than in other breeds. The collar was removed, and the seizures stopped even after medicine discontinuation.

A Cocker Spaniel owned by Barbara Trissler of El Monte, California, was not so lucky. The dog received a flea collar and within three days was dead. Since the dog had not eaten anything, since no poisons were found in its system, and since this was the first time the dog had ever worn a flea collar, Miss Trissler and her veterinarian concluded flea-collar toxicity was the cause of death.

Never, under any circumstances, use a flea collar on a puppy; the chemicals are too strong. Also, if the collar is placed on an animal that is wet or gets wet, the collar eats right through the skin on the animal's neck. I knew a Collie once that was brought to a groomer friend of mine for care. The dog had gotten wet several times, and the collar had not been removed. The groomer had to contact the vet to have the collar surgically removed because it had eaten into the animal's skin.

The alternative to the flea collar containing organic phosphates is Parid Bomb Flea Spray. I have tried many flea sprays and have found this one to be the most effective on the market and still safe for dogs and cats. To treat a cat, I saturate a cotton ball and rub it thoroughly around the animal's ears, neck, and under its front legs where the legs join the body. Then I suspend the cat by the scruff of its neck and lightly spray the rest of the body, being very careful to avoid the face. Animals do not like the spray, and I would prefer a better dispensing apparatus, but it is necessary. A good flea powder can be effective too since it contains fungus killers. Many times the dog or cat will also pick up a fungus, along with the fleas, and this will cause itching. The powder will help clear both parasites from the animal's body.

Flea tags have a very limited effect. The chemicals they contain cannot be absorbed into the bloodstream and therefore cannot be ingested by the fleas. The vapor released by these chemicals will kill the fleas around the animal's neck, where the parasites do spend much of their time. However, they can also be dangerous when dangling from a collar because the collar can then dip into the animal's drinking water. The chemical is released in water; the animal ingests it and may become dangerously ill.

Flea collars, however, are not the only carriers of toxic materials. Fly and pest strips hung in a room release gases; if the room is closed, pet birds can die from inhaling the gas. The safest method of catching flies is a fly trap. This is a simple device shaped like an angel food cake pan with legs which hold it several inches off the surface on which it is set. The bottom of the trap contains water, which should be mixed with a small amount of detergent. The top of the trap has a conelike lid, which sits over the bottom, sealing it. The bottom center has a hole in it. Place a piece of meat partially over the opening. This attracts flies, and once they enter the trap, they cannot get out; they fall into the toxic water below. This apparatus attracts the breeder flies which are looking for a place to lay their eggs. And if your animal gets at it, there are no poisonous chemicals to harm your pet.

Poisons

A final warning about poisons concerns phenol and its derivatives. Regardless of label safety claims, do not use it. Dr. Philip N. Ramsdale of Pacific Palisades, California, cites the reasons in an article titled "Household Chemicals Dangerous to Cats," published in *Cats Magazine,* July 1976, and reprinted by *Pet Pride.*

Phenol and phenol derivatives are very toxic to cats. This applies to nearly all coal-tar derivatives as well. Some compounds are more toxic than others, particularly those used as disinfectants. One of the nationally advertised household disinfectants is the Lysol line of products. I mention this because recently on major network TV commercials, Lysol spray is recommended to deodorize the cat box.

Acute phenol poisoning is not often seen in cats. Chronic or sublethal poisonings are often not diagnosed because the symptoms bear little resemblance to those of acute phenol poisoning. If, in taking the history of the case, the use of phenol or other coal-tar derivative disinfectants in housekeeping does not come to light, the diagnosis may easily be missed. Cats with obscure illnesses, one after another, following no set pattern, often have a common history of the owner's using these antiseptics in the household—sprayed "everywhere" as in the TV commercials. Once the use of these products ceases, the cat may miraculously recover. Some, however, may end up with liver and kidney damage and non-specific anemia. If enough is absorbed, it can be fatal. Pine oil is another potentially dangerous coal-tar derivative (Pinesol).

Cholorinated hydrocarbon insecticides, such as DDT, Chlorodane, and lindane, have long-lasting residual action. These affect the central nervous system and may eventually cause convulsions and death. Recovery from toxicity with central nervous symptoms is very rare.

Organic phosphate insecticides for use on or around cats are safe only if used strictly according to directions. Even then, some cats may have individual idiosyncrasy to any or all of these. DDVP (Vapona) and malathion are examples of this group.

Every veterinarian I have spoken to agrees with Dr. Ramsdale, and my experience fully substantiates his theory. Several clients have come to me with very ill cats. They had been examined by their vets and no diagnosis made or treatment found. As I communicated with the cats and "looked" into their bodies, what they told me and what I saw made me think that the cats were being poisoned somehow. Many of the cats were confined to the house; so they could not get any outdoor poisons such as snail bait (very attractive to cats) or rat poisoning (rats eat it and die, and then the cats eat the rats and get poisoned). I couldn't then and there spot the poison, but when I read Dr. Ramsdale's article, I called the owners of the cats and asked about their use of disinfectants containing phenol or phenol derivatives. Without exception, every one of them was using such a disinfectant to clean the litter pans. The poisoning was so gradual it was not detected. The cats use the pan or walk across a mopped floor. They then lick their paws to clean themselves and ingest the chemical. Gradually they get diarrhea, become lethargic, lose their appetites, and then die from the accumulation of the poison in their systems. My veterinarian says he has seen entire litters of puppies dying because the owners used a phenol disinfectant to clean the puppy boxes. To avoid exposing your pet to such danger, be sure to check the list of ingredients printed on the label of the cleansing product before you buy it, making sure that it does not contain phenol or a phenol derivative.

The safest and most effective product to use for eliminating pet odors and disinfecting floors is rubbing alcohol. There are no dangerous side-effects. I also suggest a little baking soda to kill odors in the litter clay and, of course, immediate removal of the cat stools.

One last word: Whenever you are in doubt about a pet product—anything from food to flea prevention—ask your veterinarian for advice. Sometimes even a second or third opinion may be helpful.

The industries I have mentioned have been working to resolve the problems discussed here. Hopefully, through continued research and cooperation with pet owners and veterinarians, they will lessen or entirely eliminate the undesirable side effects of some of their products. As pet owners we should collectively and cooperatively encourage these industries in their admirable efforts to improve the health and well-being of our animal friends.

PART III

MY FAITH

12

MY FAITH

Over the past seven years I have known many animal owners that have lost a dearly loved pet. Their grief is just as real and deep as that of an individual who loses a human loved one to death. I experienced this myself when my first German Shepherd, Princess Royal, died.

One night I was awakened by the sound of horrible kicking and thrashing. I flicked the light switch and stood in horror while Princess Royal writhed in a convulsion. The attack passed but was soon followed by another. I called my vet, and within minutes he met me at the animal hospital to treat her for what we both thought was food poisoning. For four days I begged God to heal my precious dog, despite the vet's eventual diagnosis of brain tumor. The case looked hopeless, he said, and yet I couldn't bring myself to say the words that would have her put to sleep.

It was 4:45 P.M. on a rainy Thursday afternoon, October 1968, when I was driving to the hospital and asked God for specific help and guidance: *Please give me a sign so that I will know your will. If she is the same or worse, I'll say those words, but if she is slightly improved, I will know you are going to heal her.* I arrived at the hospital, and the nurse told me that at 4:45 P.M. Princess Royal had suffered another attack and died. How good God was to spare me the agony of decision. He had taken her. It was his will.

Yet I felt enormous guilt at not being with Princess Royal

during her last agonizing moments, and I felt overwhelming grief and loss. I sat on my bed, sobbing, my eyes closed in pain, when God again showed his presence and purpose.

I had never had a vision; so I was unprepared for the feeling of being watched. I suddenly opened my eyes, and there sat Princess Royal, gazing up with the love she had always given me, one paw on my lap. I held her close in speechless joy. In a few minutes, without any sign of movement, she was out of my grasp. I called to her to stay, but she answered soundlessly, *I was allowed to come and tell you that I understand why you could not be with me in the hospital and that everything is all right. I came to say good-bye, but I'll see you again.* Suddenly she was gone, and I sat alone, but now with a sense of peace. I knew she was alive and with God. It was not until several months later that I learned that she could *feel* my love for her even though I was not in the hospital with her. Animals have taught me that we humans can hold onto them too tightly. At the moment I released Princess Royal with my prayer, she was free to go where God wanted her. Until then I had been holding her here in her suffering body.

The pain of her loss stayed with me for some time; so I finally decided to visit my parents in Florida. I was lying on my bed, in the home where I first got her as a puppy, when I felt the pain of missing her rush over me again and threaten to engulf me. Suddenly the ceiling opened up, and I saw her playing in the most beautiful forest, romping on a lush green carpet of grass, and then in a stream, diving under the water for rocks just as she had on our hikes in the mountains near Azusa, California. Instantly, she was standing on the bank, as beautiful as I had ever seen her, her coat dry and fluffy. She trotted over to the edge of the vision, looked down at me, and said, *We just wanted you to know that I'm all right and happy. I'm waiting for you and will see you when you get here.* Again she was gone.

After her death and these two comforting visions, I began to search the Bible for what God had to say about animals. I have found this a fascinating study through the years, and it has resulted in my personal theology.

All, except for sea, mammals are biologically much the same as humans. They also have souls, in the sense that they experience emotions as people do, and they possess a limited reasoning ability which I include in the "soul" part of the animal. The big difference between a human and an animal is in the area which is best described as the image of God. Humans possess an image of God; animals do not. Humans have a conscience or a sense of sin, a complex sexuality which involves emotion, an ability to make intricate plans for the future and to think abstractly. Animals have no sense of sin; their sexuality is only for reproduction; and they cannot plan changes in their environment. Since animals do not possess this image of God and cannot sin, I believe their souls live on. Job said, "For the soul of every living thing is in the hand of God, and the breath of all mankind" (Job 12:10, LB). The Bible does not say whether or not these creatures live on, but I believe Princess Royal is spiritually alive and whole. The Bible helped assuage my grief with these words from Revelation 21:4 : "He will wipe away all tears from their eyes, and there shall be no more death, nor sorrow, nor crying, nor pain. All of that has gone forever" (LB).

Since my ESP ability has become generally accepted, I am often asked to speak to a dead pet or a departed loved one. My answer is an emphatic *no.* I know the feeling of grief and the need to communicate, and I have just described my personal visions of Princess Royal. But I do not involve myself in the occult of any sort or in metaphysical philosophies or in anything that concerns trances, hypnosis, or anesthesia. When I became involved in subjective communication and became aware of my ESP abilities, I was eager to learn everything I could. I attended meetings in which current cult philosophies were taught or demonstrated. I studied the works of J. Allen Boone, listened to California psychic Fred Kimble lecture, and investigated the teachings of Edgar Cayce. I felt something was wrong. When a person is in a trance or under hypnosis or anesthesia, he or she is open to many influences and is dangerously vulnerable. When I knew of my abilities, I prayed to God and told him I would accept his work, but only if I could do it under complete control, wide awake and aware of my environment at all times. He has honored my request. The only spirit I communicate with is the Holy Spirit, and that is through private and conscious prayer.

I *do* believe in God and in his Son, Jesus Christ, and in the Spirit of God. I also believe in the Bible as the Word of God. I do *not* believe in modern philosophies which teach that all humans have a bit of the God image in them, as Jesus did, and that he was a human being just like the rest of us. I believe in his divinity and our nondivinity. I also believe that my prayers and God's guidance have helped me to use my gift for the good of animals and humankind or, in other words, to do his work.

God created the animals and then created man in his image to have dominion over the animals. Man sinned and when sin entered the world, man suffered; so did the animals. But I believe that when the second coming of Christ takes place, the animals will also be released from suffering. I take comfort from Romans 8:18–22:

> Yet what we suffer now is nothing compared to the glory he will give us later. For all creation is waiting patiently and hopefully for that future day when God will resurrect his children. For on that day thorns and thistles, sin, death, and decay—the things that overcame the world against its will at God's command—will all disappear, and the world around us will share in the glorious freedom from sin which God's children enjoy.
>
> For we know that even the things of nature, like animals and plants, suffer in sickness and death as they await this great event (LB).

I am also asked about reincarnation in animals because apparently the theory of reincarnation gives some logical answers to previously unanswered questions. "Before every man there lies a wide and pleasant road that seems right but ends in death"

(Prov. 14:12, LB). In all my experiences with animals, I have never seen any evidence of reincarnation. I have asked many animals about this. One example should suffice. A dog owner, who believed in reincarnation and thought his dog was a former dog of his, made the living dog so confused he didn't know who he was and also felt he didn't have the right to be himself. The dog communicated to me that all he knew was what was being mentally expected of him, namely the behavior of the dead dog. So he responded accordingly, in utter confusion. To return to the Bible, "We must be born again" (see John 3:3). This is *spiritual* rebirth that takes place at the moment we accept Jesus Christ as personal Savior; this is not reincarnation.

Communicating with and treating animals, and also studying the Bible, I saw animals all through the Bible playing a significant role in human life and in God's scheme of things. Animals are described as instruments of God, either to minister to, feed, or bring judgment on the sins of men and women. (See 1 Kings 21:21 and 2 Kings 2:24: The boys that worshiped other gods mocked the servant of the true God.)

Many people say they do not believe the biblical account of creation, but I do. Reading Genesis, I am convinced that there was complete harmony between humans and animals until after the flood and that all animals were herbivores until then. Adam and Eve communicated with the animals in the Garden of Eden. Why else would Eve talk to the serpent and respect its ideas as reasonable if it were not a usual experience to communicate with animals? I believe that all communication in the garden was nonverbal and harmonious; it is therefore not surprising that later the animals obeyed Noah's command to enter the ark and live there peacefully. After the waters receded from the flood, God told Noah that from that day on God was delivering every beast of the field and every fowl of the air to be food for humans, and he placed the fear of humans in animals. I have discovered this natural fear in the wild animals with whom I've communicated. They don't usually run away, but they let me know they do not trust me fully because I am a human being. I believe this true communication between humans and animals existed on its nonverbal level until the Tower of Babel. I also believe God made each creature reproduce after its own kind so that even today, when scientists attempt to cross-breed animal species, the offspring is sterile.

In Numbers 22:21–35 we read about Balaam going along the trail on his donkey. Three times Balaam beat the donkey, and then God allowed her to verbalize what she was thinking. *Why do you hit me? Have I ever disobeyed you before?* Then God opened Balaam's eyes—his mind's eye—and Balaam saw the angel standing in the way. I believe the donkey was speaking its own mind because I have found that animals often see nonphysical beings.

Elijah was fed by the ravens by the brook (see 1 Kings 17:1–6), the perfect precedent to Mrs. Ely Buffin's experience with the eagle.

Throughout the Old Testament lambs were used as sacrifices for the sins of men and women as a symbol of the ultimate sacrifice of the perfect Lamb of God, Jesus Christ, who died for the sins of all humankind. The unbroken donkey colt submitted to being ridden by Jesus when Jesus made his triumphal entrance into Jerusalem just before his crucifixion.

As you search the Scriptures, you will see the harmony God intended for human beings and beasts in his creation. "No mere man has ever seen, heard or even imagined what wonderful things God has ready for those who love the Lord" (1 Cor. 2:9, LB). Man's vision is narrow, but God's is limitless. "Are not two sparrows sold for a farthing? and one of them shall not fall on the ground without your Father" (Matt. 10:29, KJV). Or, translated literally from the original Greek, ". . . shall not fall on the ground without the Father being right there with it as it falls."

I believe God is trying to tell us that he made this earth and the creatures on it, that he cares what happens to his creation, and that he wants to help humankind toward a fuller and richer relationship with the animals of his earthly kingdom.

I also believe God sees the importance and the emotional necessity of animals in the lives of humans, especially when there is a void in those lives. But a person who tries to treat an animal as his or her lost child or lost loved one puts a terrible emotional strain on the animal and may make it physically ill. It will certainly make both owner and animal neurotic. My animals are happy and well adjusted because they are loved as my substitute children and companions, but they are just *substitutes* and are still treated as animals.

As I began to understand God's concern for animals, I asked him to show me what he would have me do with my abilities. This book is my answer to what I believe is his answer.